Cambridge star atlas 2000.0

Wil Tirion

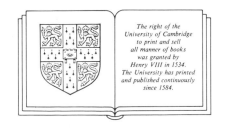
CAMBRIDGE UNIVERSITY PRESS

Cambridge
New York Port Chester
Melbourne Sydney

Published by the Press Syndicate of the University of Cambridge
The Pitt Building, Trumpington Street, Cambridge CB2 1RP
40 West 20th Street, New York, NY 10011–4211, USA
10 Stamford Road, Oakleigh, Melbourne 3166, Australia

First published 1991

Printed in Great Britain at the University Press, Cambridge

British Library cataloguing in publication data

Tirion, Wil
Cambridge star atlas 2000.0.
1. Stars – Maps, atlases
I. Title
523.8'903

Library of Congress cataloguing in publication data
Tirion, Wil.
Cambridge star atlas 2000.0 / Wil Tirion
p.   cm.
ISBN 0 521 26322 0
1. Stars – Atlases. I Title.
GB65.W5   1991
523.8'022'3–dc20 – 89–38035 CIP

ISBN 0 521 26322 0 hardback

Cambridge star atlas 2000.0

Contents

Preface

Anyone who looks up at the starry sky at night and wonders how to find a way among all those stars will need some kind of sky-guide or atlas. But very different needs must be met. The casual star-gazer will first want to learn what can be seen with the unaided eye; the names of the stars, the constellations and where or when to look for Orion, the Great Bear, or for Andromeda. The more advanced observer, with access to a good pair of binoculars or a small telescope, wants to know more. Where is the Whirlpool Galaxy, where the North-America Nebula, or the globular cluster M13?

The *Cambridge Star Atlas 2000.0* offers help for both. It includes a series of twelve monthly sky-maps, designed to be of use for almost anywhere on earth together with a series of twenty detailed star-charts, covering the whole heavens, with all stars visible to the naked eye under good circumstances. The star-charts also show a wealth of star-clusters, nebulae and galaxies. Some of these can be seen without optical help, but for most a small or average-size telescope is needed. Accompanying the charts are tables, compiled by Patrick Moore, giving more information about the interesting objects on each chart.

Finally, the atlas includes a series of six all-sky maps in which the northern and the southern hemispheres are set out side by side, showing the distribution of stars, open and globular clusters, planetary and diffuse nebulae and galaxies.

Happy stargazing!

WIL TIRION

The monthly sky maps

Most people know one or two constellations. They probably know what the Great Bear looks like. But why is it always in a different position in the sky? And why is it not possible to find Orion during a summer's night? This changing aspect of the sky is often confusing to the casual star-gazer. So, the first thing one has to learn is how the sky moves.

It is important to know that the star-patterns themselves are not changing, at least not in a single human life-span. It is only over a period of centuries that the positions of some nearby stars change in a way that can be detected with the unaided eye. Now all those groupings of stars and constellations can be regarded as being fixed to a huge imaginary sphere, with the earth placed in the centre. No matter where on earth we are, we can always see just one half of this sphere. So it is not hard to understand that, when we move to another part of the earth, the visible part of the sphere also changes. If we stand at the North Pole we will only see the northern half of the heavens, while at the South Pole we will see only the southern half. But it is not all that simple. Two more influences affect the aspect of the sky. First there is the daily rotation of the earth around its axis, making the sun come up in the east and set in the west. The same thing happens with the other objects in the sky. In fact, the heavenly sphere seems to rotate around an axis that is an extension of the earth's axis. Then there is the orbital movement of the earth, making the appearance of the sky change over the seasons. Look at one constellation, let us say Orion, at midnight on 1 January and take note of its position. Then look every following night at about the same time and you will notice that the stars reach the same position a few minutes earlier every night. One month later, on 1 February, Orion will already be in that place at 10 PM. This will continue, slowly changing the look of the night sky, until one year later the earth has reached the same place in its orbit again and Orion will be back in its original position at midnight. It is interesting to realize that at the North Pole we always see the same part of the sky, since the earth's rotation will only cause us to turn around our own axis. The sky turns around the point right above us, the celestial North Pole, and stars only move parallel to the horizon, they do not rise or set. At the South Pole the same thing would happen. And since the axis of our planet does not change its position in relation to the stars when it moves around the sun the visible sky remains the same all year long. In fact of course, all six months long, the other six months will have daylight. On the other hand, on the equator we will see a

1

quite different situation. The celestial Equator, or the line where the plane of the earth's Equator cuts the celestial sphere, is running from the east, right overhead to the west, and all stars rise and set, and during the year it is possible to see the entire sky. Finally, in the intermediate areas the situation is more complicated. Some stars rise and set, others are always invisible, and again other stars never drop below the horizon; they are 'circumpolar' (Fig. 1).

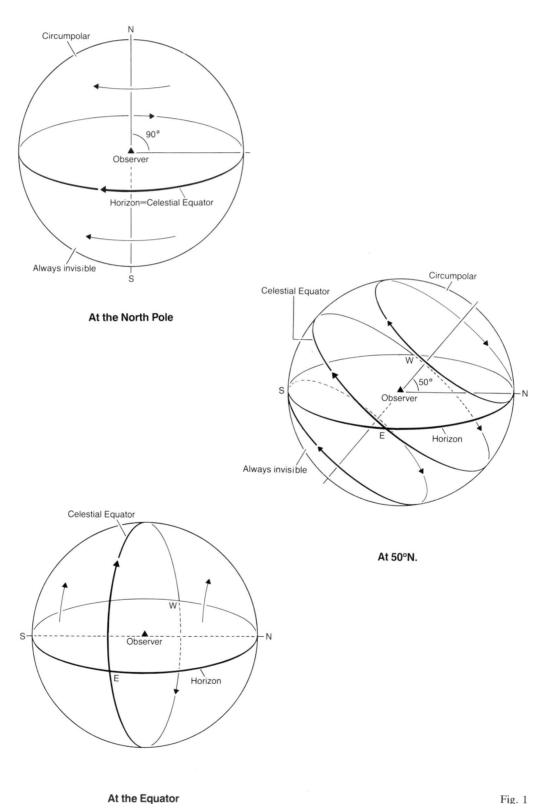

At the North Pole

At 50°N.

At the Equator

Fig. 1

2

The twelve monthly sky maps are constructed on a Transverse Mercator-Projection. The advantage of this projection is the possibility of showing a number of different horizons, northern as well as southern, as simple straight lines. So, for a wide range of places on earth, between 60° N and 40° S you can draw in your own horizon. The disadvantage is that the east and west points of the horizon cannot be shown.

Choosing the right map

There are twelve maps, one for each month. The maps are for 11 PM on the first day of the given month, 10 PM on the 15th and 9 PM on the first day of the following month. These times are given at the bottom (right) of each map. But the maps can be used over a range of dates and times, as is shown in Table A. When local Summer Time (Daylight Saving Time) is used, one hour should be added to the times given in the table (or one hour subtracted from the time on your watch). Once the right map is found, use a straight-edge to draw the horizon you need (dependent on your earth latitude). This can simply be done by connecting the markings along the left and right margins of the map. Two strips of paper can also be used to cover the stars below your horizon. In the north the labels are upside-down. This is because the map has to be turned when looking to the north. The star's brightnesses are given at the bottom left. The meaning of the stellar 'magnitudes' is explained in the introduction to the main star charts.

Once you have learned how to use the maps and find the constellations you will want to know more of what can be seen up there. The grey numbers overprinted on the monthly maps indicate which of the more detailed star charts to use.

Table A. Selecting the monthly sky maps

Month	Day	5 PM	6 PM	7 PM	8 PM	9 PM	10 PM	11 PM	Midnight	1 AM	2 AM	3 AM	4 AM	5 AM	6 AM	7 AM
January	1	Oct		Nov		Dec		Jan		Feb		Mar		Apr		May
January	15		Nov		Dec		Jan		Feb		Mar		Apr		May	
February	1	Nov		Dec		Jan		Feb		Mar		Apr		May		Jun
February	15		Dec		Jan		Feb		Mar		Apr		May		Jun	
March	1	Dec		Jan		Feb		Mar		Apr		May		Jun		Jul
March	15		Jan		Feb		Mar		Apr		May		Jun		Jul	
April	1	Jan		Feb		Mar		Apr		May		Jun		Jul		Aug
April	15		Feb		Mar		Apr		May		Jun		Jul		Aug	
May	1	Feb		Mar		Apr		May		Jun		Jul		Aug		Sep
May	15		Mar		Apr		May		Jun		Jul		Aug		Sep	
June	1	Mar		Apr		May		Jun		Jul		Aug		Sep		Oct
June	15		Apr		May		Jun		Jul		Aug		Sep		Oct	
July	1	Apr		May		Jun		Jul		Aug		Sep		Oct		Nov
July	15		May		Jun		Jul		Aug		Sep		Oct		Nov	
August	1	May		Jun		Jul		Aug		Sep		Oct		Nov		Dec
August	15		Jun		Jul		Aug		Sep		Oct		Nov		Dec	
September	1	Jun		Jul		Aug		Sep		Oct		Nov		Dec		Jan
September	15		Jul		Aug		Sep		Oct		Nov		Dec		Jan	
October	1	Jul		Aug		Sep		Oct		Nov		Dec		Jan		Feb
October	15		Aug		Sep		Oct		Nov		Dec		Jan		Feb	
November	1	Aug		Sep		Oct		Nov		Dec		Jan		Feb		Mar
November	15		Sep		Oct		Nov		Dec		Jan		Feb		Mar	
December	1	Sep		Oct		Nov		Dec		Jan		Feb		Mar		Apr
December	15		Oct		Nov		Dec		Jan		Feb		Mar		Apr	

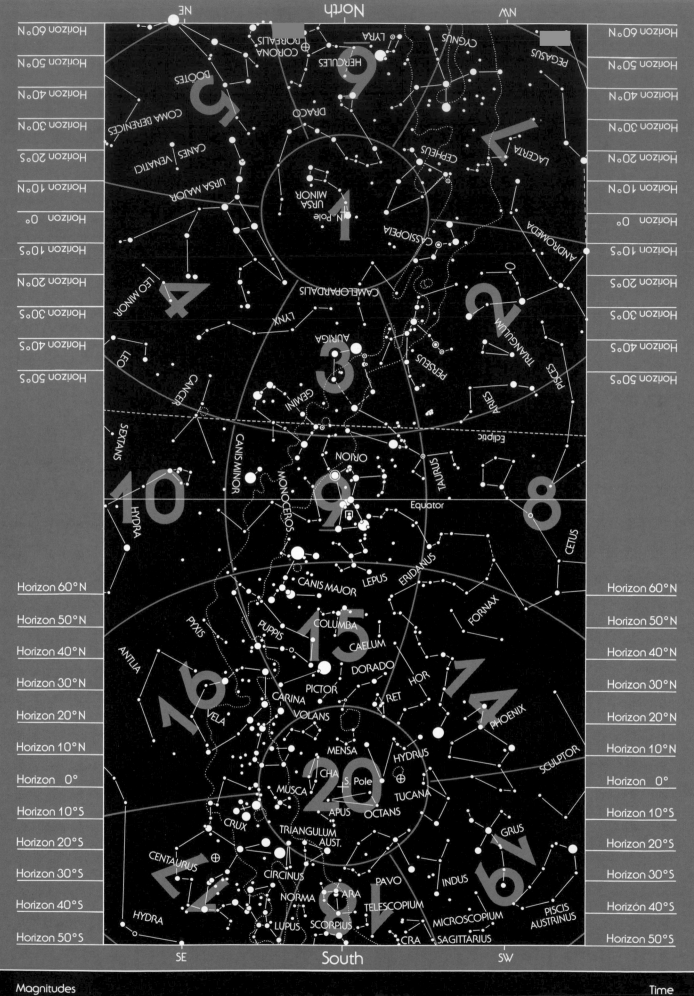

JANUARY

Magnitudes
● ● ● ● · ·
−1 0 1 2 3 4-5

	Time
January 1	11 PM
January 15	10 PM
February 1	9 PM

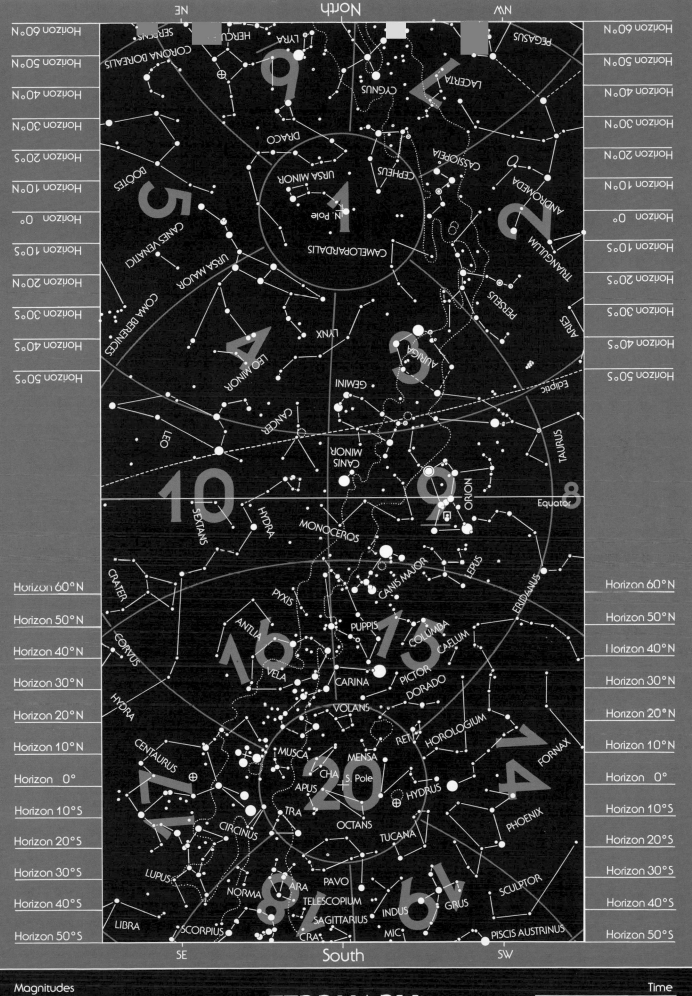

FEBRUARY

	Time
February 1	11 PM
February 15	**10 PM**
March 1	9 PM

Magnitudes

−1   0   1   2   3   4-5

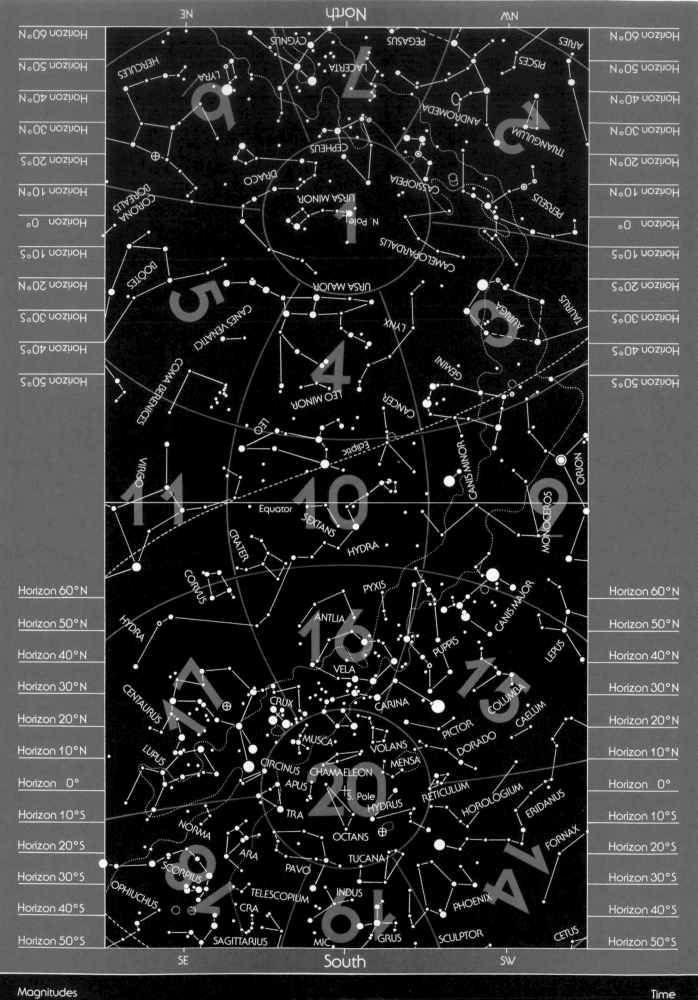

MARCH

Magnitudes						Time	
●	●	●	•	·	·	March 1	11 PM
−1	0	1	2	3	4–5	March 15	10 PM
						April 1	9 PM

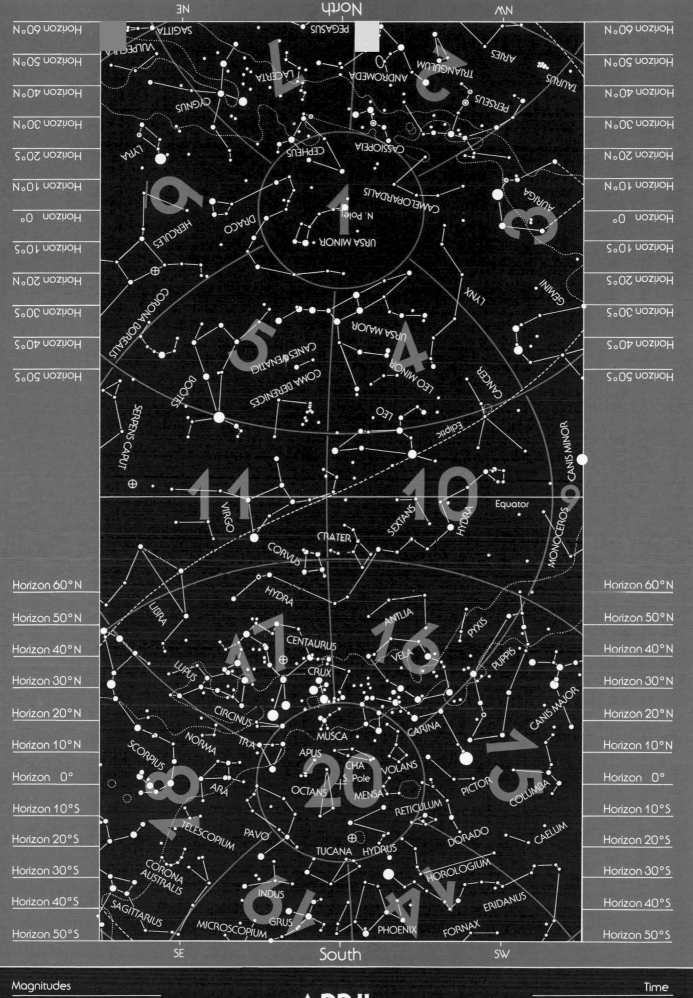

APRIL

Magnitudes

● ● ● · · ·
-1 0 1 2 3 4-5

	Time
April 1	11 PM
April 15	**10 PM**
May 1	9 PM

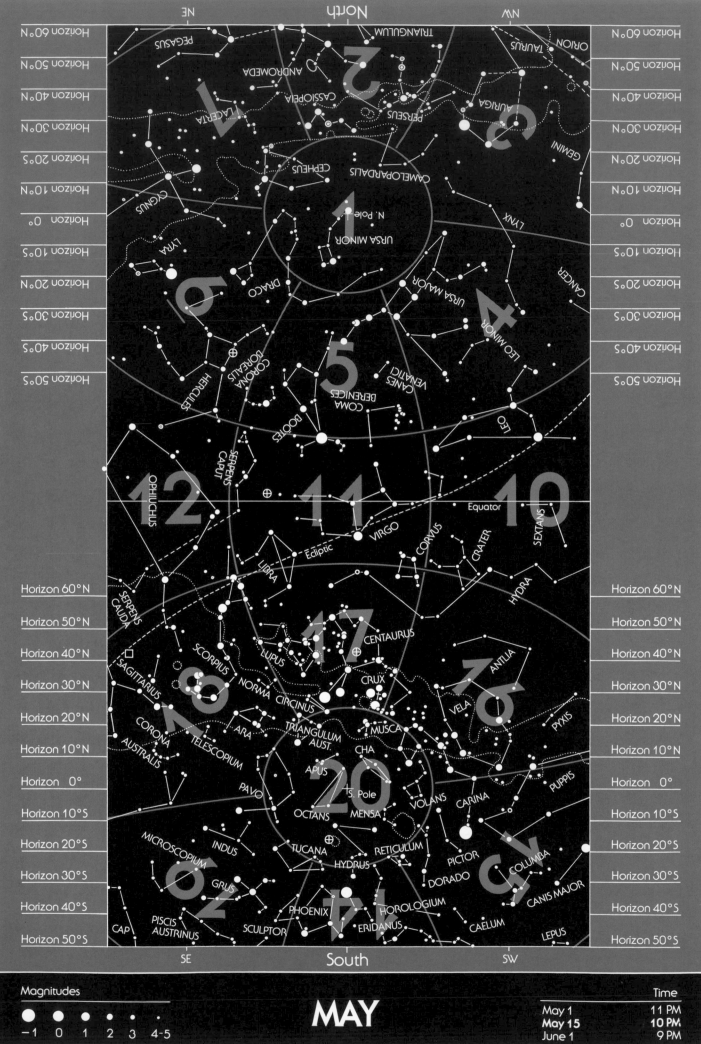

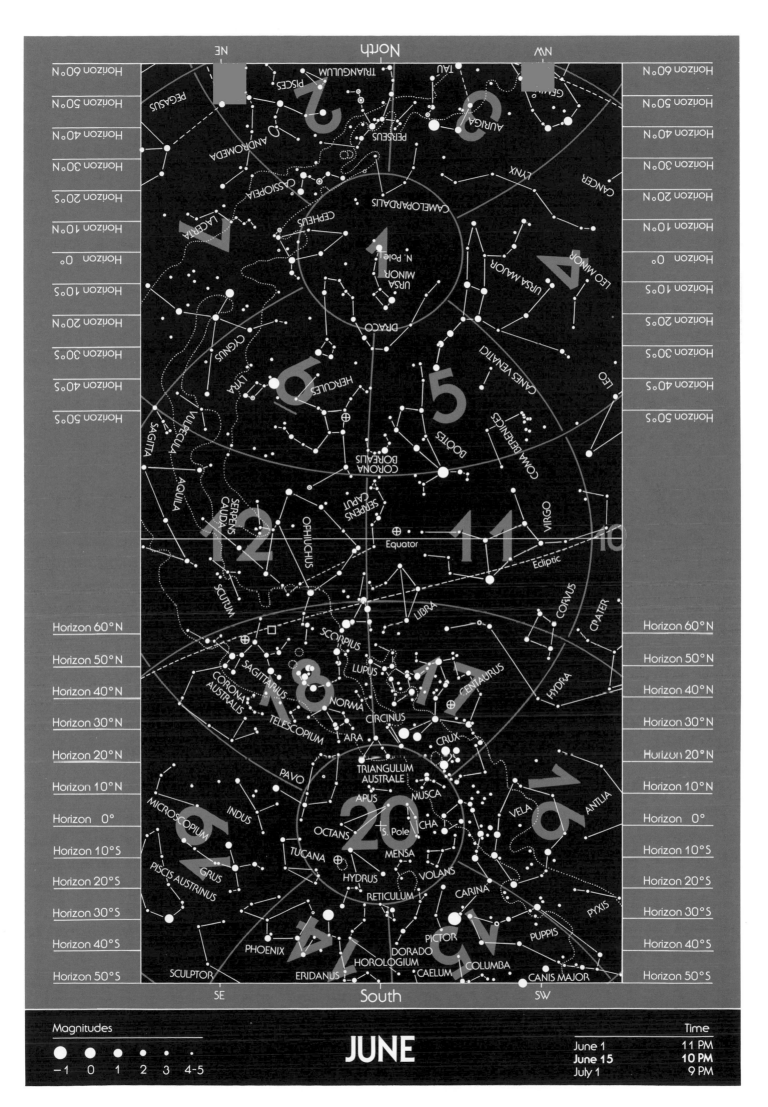

JUNE

	Time
June 1	11 PM
June 15	**10 PM**
July 1	9 PM

Magnitudes

−1   0   1   2   3   4–5

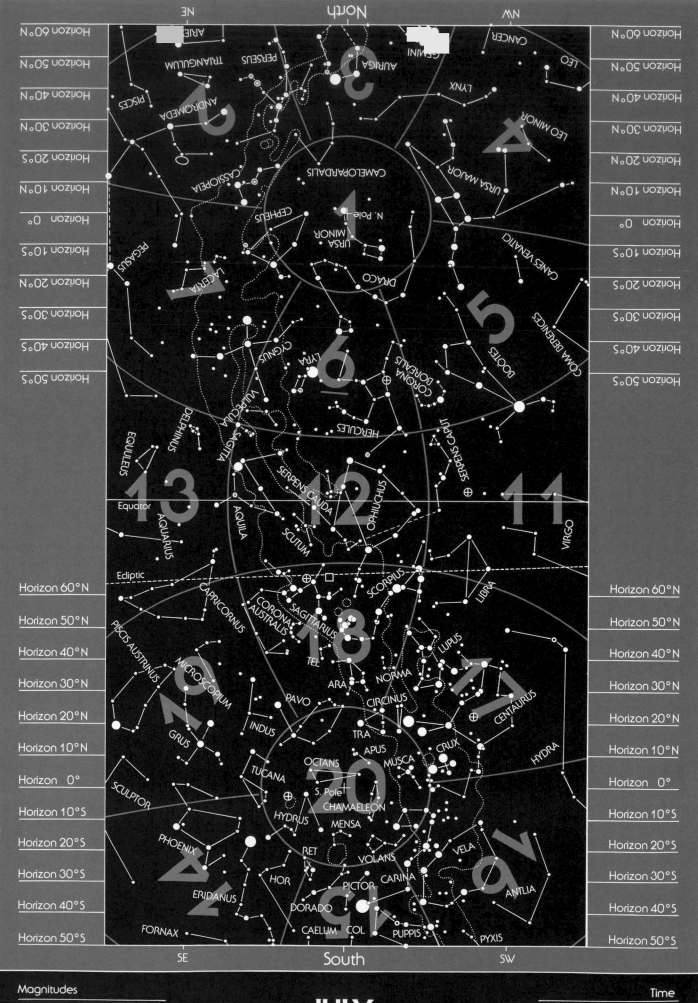

JULY

Magnitudes

● ● ● ● · ·
-1 0 1 2 3 4-5

	Time
July 1	11 PM
July 15	**10 PM**
August 1	9 PM

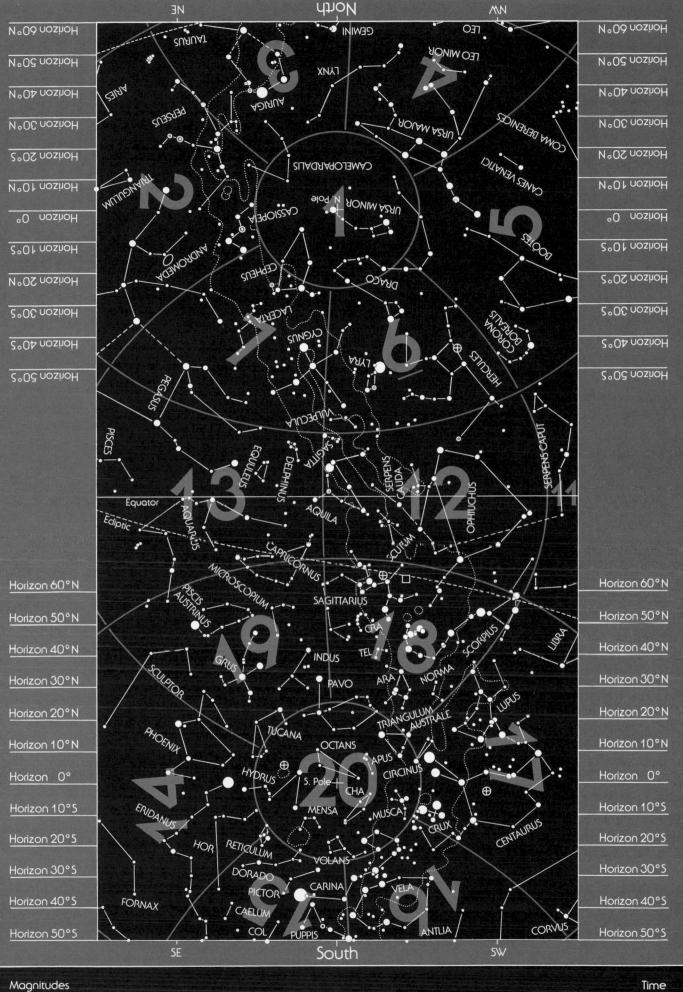

AUGUST

	Time
August 1	11 PM
August 15	**10 PM**
September 1	9 PM

Magnitudes

−1 0 1 2 3 4-5

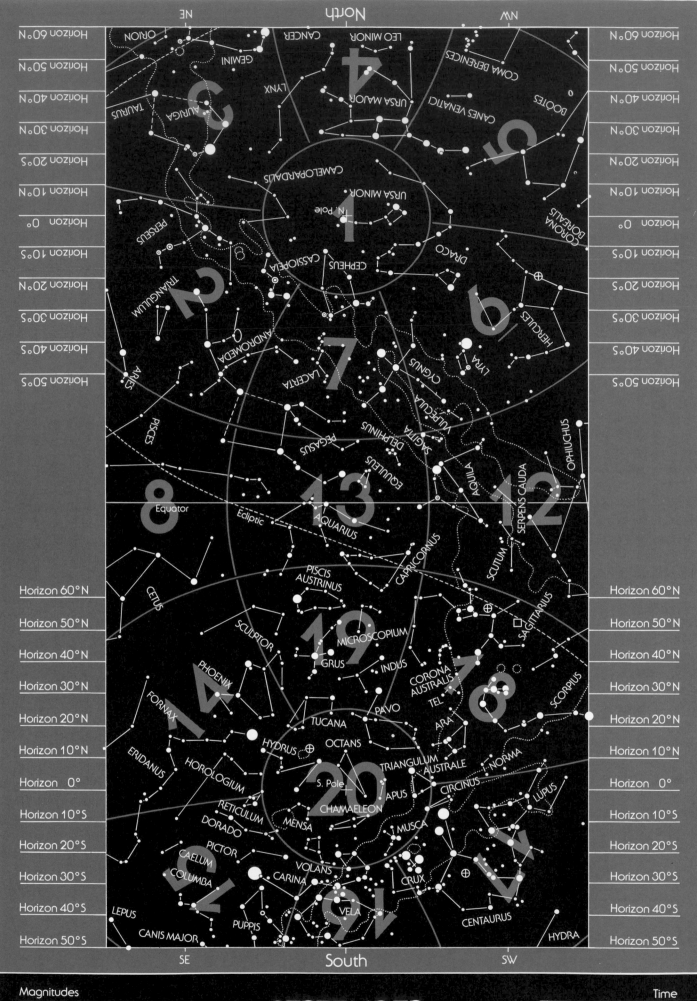

SEPTEMBER

Magnitudes

-1　0　1　2　3　4-5

Time	
September 1	11 PM
September 15	**10 PM**
October 1	9 PM

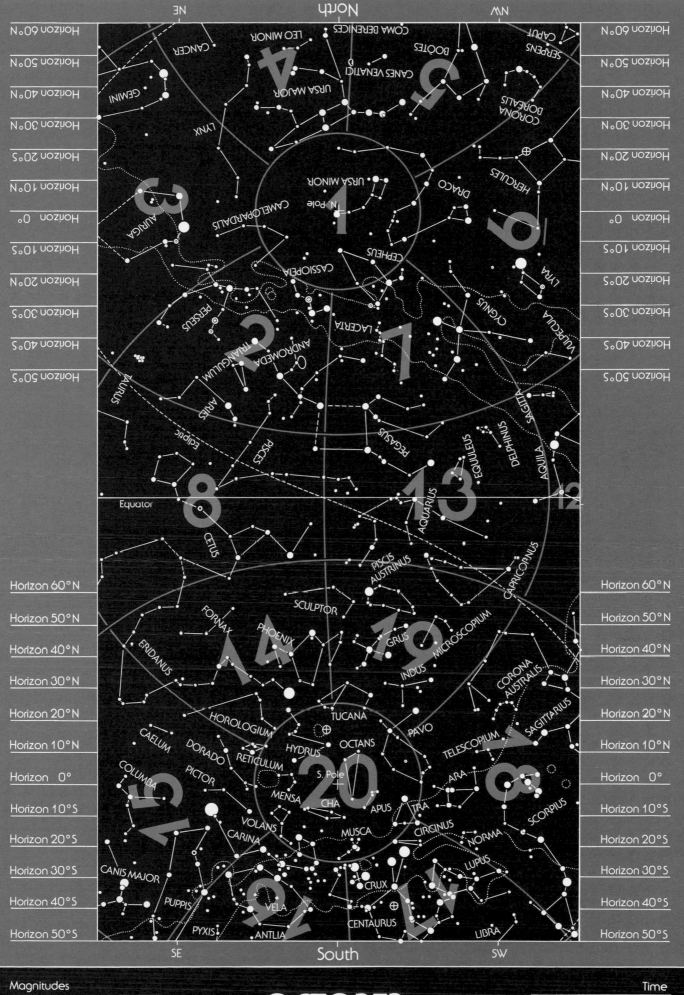

OCTOBER

Magnitudes

● ● ● ● · ·
−1   0   1   2   3   4-5

	Time
October 1	11 PM
October 15	**10 PM**
November 1	9 PM

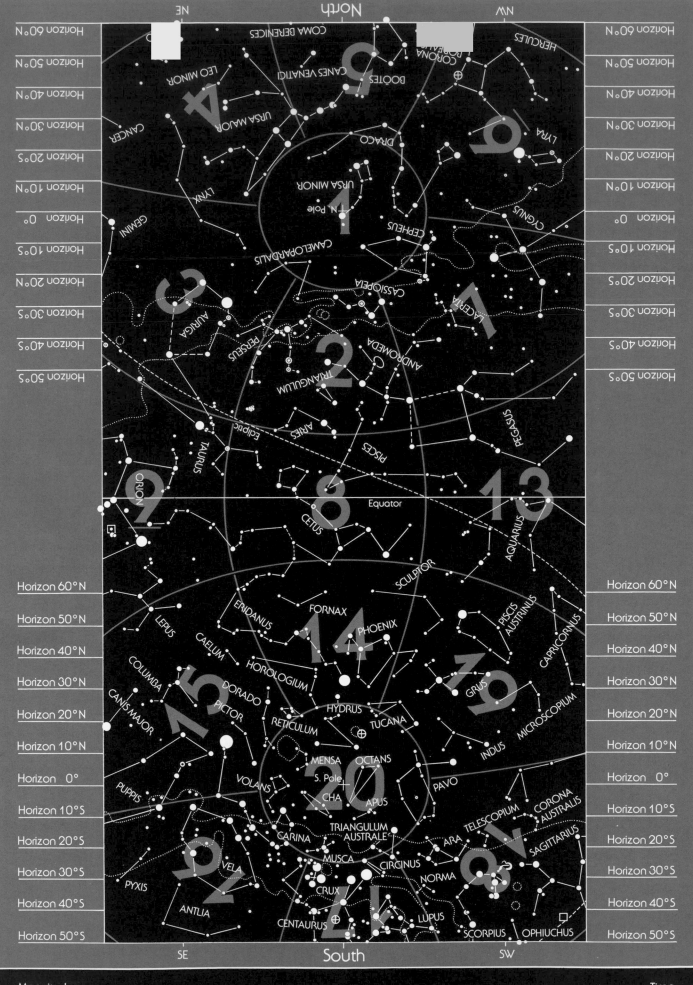

Magnitudes

● ● ● ● · · ·
-1 0 1 2 3 4-5

NOVEMBER

Time

November 1   11 PM
November 15   10 PM
December 1   9 PM

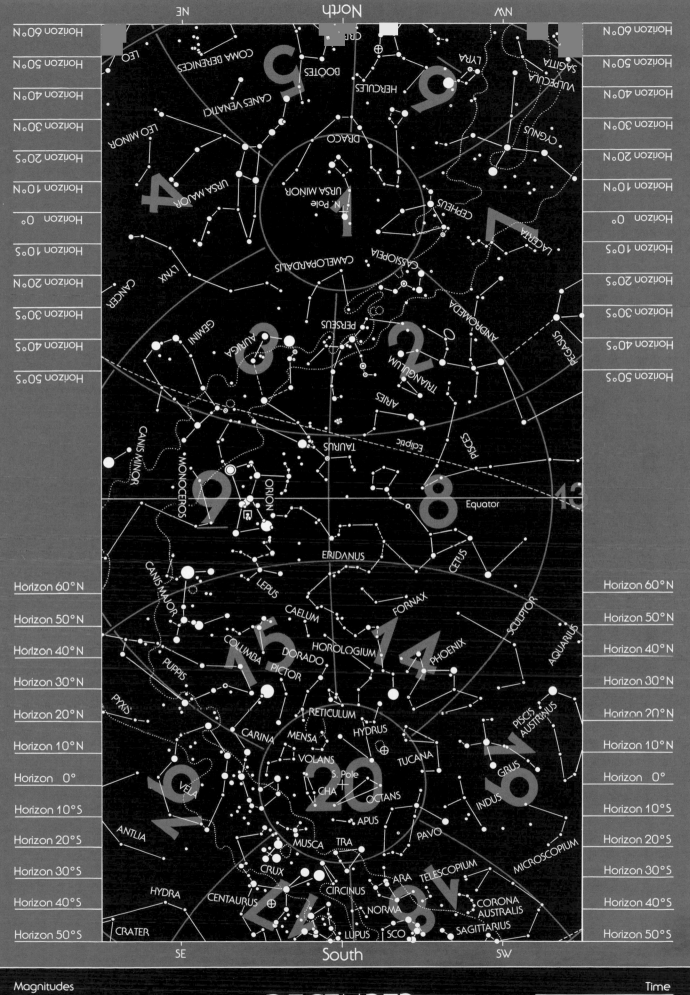

DECEMBER

Magnitudes	Time
● ● ● ● · ·	December 1 — 11 PM
−1 0 1 2 3 4-5	December 15 — 10 PM
	January 1 — 9 PM

The star charts

The main star charts divide the sky into 20 parts. The actual chart areas are shown in Table B. There is a generous overlap between the charts, so most of the constellations are shown complete on at least one chart. The positions of stars and objects are for the year 2000, or to be more precise, the epoch is 2000.0, the extra 0 is a decimal and it means 1 January. (2000.5 would be 1 June.) The positions are plotted against a grid of right ascension (RA) and declination (Dec) comparable with longitude and latitude on the Earth's globe. Right ascension is reckoned in hours, minutes and seconds from 0h to 24h, from west to east along the Equator. Declination represents the angular distance between an object and the celestial Equator, + for objects north and − for those south of the Equator (Fig. 2). The charts' projections have been carefully chosen to show the star-patterns with the least possible distortion and to make it easy to measure positions from the maps. The projections used are azimuthal equidistant (the polar charts, 1 and 20), secant cylindrical (the equatorial band, charts 8 to 13) and the secant conic projection (the intermediate areas). The advantage of all three

Table B. Areas covered by the star charts (overlap not included)

Declination zone	Chart number	Right ascension
+90° / +70°	1	0h / 24h (0h)
+70° / +20°	2	0h / 4h
	3	4h / 8h
	4	8h / 12h
	5	12h / 16h
	6	16h / 20h
	7	20h / 24h (0h)
+20° / −20°	8	0h / 4h
	9	4h / 8h
	10	8h / 12h
	11	12h / 16h
	12	16h / 20h
	13	20h / 24h (0h)
−20° / −70°	14	0h / 4h
	15	4h / 8h
	16	8h / 12h
	17	12h / 16h
	18	16h / 20h
	19	20h / 24h (0h)
−70° / −90°	20	0h / 24h (0h)

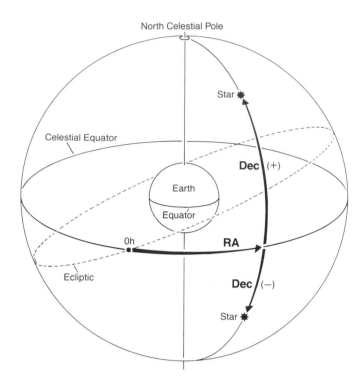

North Celestial Pole

Star

Celestial Equator

Dec (+)

Earth

Equator

0h

RA

Ecliptic

Dec (−)

Star

Fig. 2

projections is that all hour circles (lines of equal RA; running north/south from pole to pole, like the meridians on earth) are shown as straight lines, and all parallels (lines of equal declination, parallel to the Equator) are equally spaced everywhere on the charts, so positions can easily be measured with a simple ruler. The tick marks along the chart borders and on the central hour circle on the conic maps, will be helpful.

Magnitudes

The word magnitude usually refers to the apparent brightness of a star or an object. Traditionally, stars visible to the naked eye were binned into six groups of brightnesses. The most prominent stars in the sky were called first magnitude stars, the ones slightly fainter (e.g. the Pole Star) second magnitude and so on. The faintest visible to the eye were magnitude six. Nowadays astronomers are able to measure the brightness very accurately. In modern catalogues you will find magnitude to two places of decimals and the scale now has a logarithmic footing. A difference of five magnitude is defined as 100. Consequently one magnitude represents a difference of 2.5 times (or to be precise 2.512, the fifth root of 100). On this scale several stars turned out to be brighter than 1, so the scale was extended to 0, but even that was not enough. A few stars were still brighter and were given a negative magnitude. The brightest star in the sky, Sirius has a magnitude of −1.46.

On the atlas the brightness of a star is rounded to the nearest whole magnitude. Stars between 0.51 and 1.50 are magnitude 1, between 1.51 and 2.50 magnitude 2 and so on. Positions and magnitudes are taken from the *Bright Star Catalogue* by Dorrit Hoffleit (4th edition, 1982). All 9096 stars from the catalogue are plotted. Stars having a magnitude of 6.5 or brighter in the *Smithsonian Astrophysical Observatory Star Catalog* and not included in the *Bright Star Catalogue* are also added, bringing the total of plotted stars to ±9500. Furthermore, the atlas shows 866 non-stellar objects. Stars are divided into spectral classes, according to differences in their spectra. These differences are related to the surface temperature of the stars. The hottest, bluish stars are the classes O, B and A, the slightly cooler stars, white, are class F, yellow class G, orange class K and the coolest, the red stars are class M. To remember this rather illogical sequence of letters, remember the sentence: Oh, Be A Fine Girl, Kiss Me!, or if you prefer: Oh, Be A Fine Guy, Kiss Me!

Star names

Many of the brighter stars have proper names, of Latin, Greek or Arabic origin, such as Regulus, Altair and Betelgeuse. Only the brightest stars are still referred to by these old names. Most stars now bear designations of numbers and Greek letters. The German celestial cartographer Johann Bayer introduced the Greek letters. In general, the brightest star in a constellation received the first letter of the Greek alphabet, alpha (α), the second letter, beta (β) was given to the second brightest and so on, although there are many obvious exceptions to this rule. Another way to identify stars is by numbers. In each constellation the stars are numbered in order of RA. These numbers are usually referred to as Flamsteed numbers. Most of the brighter stars have both a Greek letter and a number. On the star charts you will find all Greek letters and in addition, the Flamsteed numbers for those stars not having a Greek letter. For stars of magnitude 1 and brighter, the proper names are also given, as is done for a few well-known second magnitude stars; Algol, Mira, Castor and Polaris.

Whenever a star is referred to by Greek letter or by number, this is followed by the genitive form of the Latin constellation name, or its official abbreviation. So the star Deneb, in the Swan (Latin: *Cygnus*) can also be called Alpha (α) Cygni or 50 Cygni (α Cyg, 50 Cyg). The Greek alphabet is given in Table C and Table D gives the names of the constellations, the genitive form, the official abbreviation and the common English name.

Variable stars have a quite different nomenclature. Some have the regular star identification like Algol (β Per) or Mira (o Cet) but most are labelled in a special way; by roman letters starting with R, then S, T etc. to Z. Then RR, RS, RT . . . RZ, next SS, ST, and so on up to ZZ. After these 54 the naming continues with AA, AB, AC . . . AZ, then BB, BD and so on again up to QZ. Now we have a total of 334. The next variable found in that constellation is called V335, then V336 and so on. These designations are also followed by the Constellation name, as with the Greek letters and the Flamsteed numbers.

Constellations

Most of the constellations we know originate from Mesopotamian traditions and from Greek mythology, but over the centuries many other constellations have been added to the classical ones, especially in the southern sky. Several of these new constellations only had a short life, while others have survived. In 1930 the International Astronomical Union finally adopted a list of 88 official constellations and the boundaries were also delimited once and for all. On our charts these official boundaries are drawn in as broken lines. The final column of table D gives the charts where you can find each of these constellations.

Table C. The Greek alphabet.

α	Alpha	ν	Nu
β	Beta	ξ	Xi
γ	Gamma	o	Omicron
δ	Delta	π	Pi
ε	Epsilon	ϱ	Rho
ζ	Zeta	σ	Sigma
η	Eta	τ	Tau
θ	Theta	υ	Upsilon
ι	Iota	φ	Phi
$\varkappa$	Kappa	χ	Chi
λ	Lambda	ψ	Psi
μ	Mu	ω	Omega

Table D. List of constellations.

Name	Genitive	Abbreviation	Common name	Chart number(s)		
Andromeda	Andromedae	And	Andromeda	2		
Antlia	Antliae	Ant	Air Pump	16		
Apus	Apodis	Aps	Bird of Paradise	20		
Aquarius	Aquarii	Aqr	Water Carrier	13		
Aquila	Aquilae	Aql	Eagle	12		
Ara	Arae	Ara	Altar	18		
Aries	Arietis	Ari	Ram	8	2	
Auriga	Aurigae	Aur	Charioteer	3		
Boötes	Boötis	Boo	Herdsman	5	11	
Caelum	Caeli	Cae	Engraving Tool	15		
Camelopardalis	Camelopardalis	Cam	Giraffe	1	3	
Cancer	Cancri	Cnc	Crab	10	4	
Canes Venatici	Canum Venaticorum	CVn	Hunting Dogs	5		
Canis Major	Canis Majoris	CMa	Greater Dog	9	15	
Canis Minor	Canis Minoris	CMi	Lesser Dog	9		
Capricornus	Capricorni	Cap	Sea Goat	13		
Carina	Carinae	Car	Keel	16	20	
Cassiopeia	Cassiopeiae	Cas	Cassiopeia	1	2	
Centaurus	Centauri	Cen	Centaur	17		
Cepheus	Cephei	Cep	Cepheus	1		
Cetus	Ceti	Cet	Whale	8		
Chamaeleon	Chamaeleonis	Cha	Chameleon	20		
Circinus	Circini	Cir	Pair of Compasses	17	20	
Columba	Columbae	Col	Dove	15		
Coma Berenices	Coma Berenicis	Com	Berenice's Hair	5	11	
Corona Australis	Coronae Australis	CrA	Southern Crown	18		
Corona Borealis	Coronae Borealis	CrB	Northern Crown	5	6	
Corvus	Corvi	Crv	Crow	11		
Crater	Crateris	Crt	Cup	10		
Crux	Crucis	Cru	Southern Cross	16	17	20
Cygnus	Cygni	Cyg	Swan	7		
Delphinus	Delphini	Del	Dolphin	13		
Dorado	Doradus	Dor	Goldfish	15		
Draco	Draconis	Dra	Dragon	1	6	
Equuleus	Equulei	Equ	Little Horse	13		
Eridanus	Eridani	Eri	River Eridanus	8	9	14
Fornax	Fornacis	For	Furnace	14		
Gemini	Geminorum	Gem	Twins	3	9	
Grus	Gruis	Gru	Crane	19		
Hercules	Herculis	Her	Hercules	6	12	
Horologium	Horologii	Hor	Pendulum Clock	14		
Hydra	Hydrae	Hya	Water Snake	10	16	17
Hydrus	Hydri	Hyi	Lesser Water Snake	20		
Indus	Indi	Ind	Indian	19	20	
Lacerta	Lacertae	Lac	Lizard	7		
Leo	Leonis	Leo	Lion	10	4	
Leo Minor	Leonis Minoris	LMi	Lesser Lion	4		
Lepus	Leporis	Lep	Hare	9		
Libra	Librae	Lib	Scales	11	17	
Lupus	Lupi	Lup	Wolf	17	18	
Lynx	Lyncis	Lyn	Lynx	4		
Lyra	Lyrae	Lyr	Lyre	6		
Mensa	Mensae	Men	Table Mountain	20		
Microscopium	Microscopii	Mic	Microscope	19		
Monoceros	Monocerotis	Mon	Unicorn	9		
Musca	Muscae	Mus	Fly	20		
Norma	Normae	Nor	Level	17	18	
Octans	Octantis	Oct	Octant	20		
Ophiuchus	Ophiuchi	Oph	Serpent Holder	12	18	

Orion	Orionis	Ori	Orion, the Hunter	9		
Pavo	Pavonis	Pav	Peacock	20	18	
Pegasus	Pegasi	Peg	Pegasus	13	7	
Perseus	Persei	Per	Perseus	2	3	
Phoenix	Phoenicis	Phe	Phoenix	14		
Pictor	Pictoris	Pic	Painter's Easel	15		
Pisces	Piscium	Psc	Fishes	8	13	2
Piscis Austrinus	Piscis Austrini	PsA	Southern Fish	19		
Puppis	Puppis	Pup	Stern	15	9	
Pyxis	Pyxidis	Pyx	Mariner's Compass	16		
Reticulum	Reticuli	Ret	Net	14	15	
Sagitta	Sagittae	Sge	Arrow	6	12	
Sagittarius	Sagittarii	Sgr	Archer	18	12	
Scorpius	Scorpii	Sco	Scorpion	18	12	
Sculptor	Sculptoris	Scl	Sculptor	14	19	
Scutum	Scuti	Sct	Shield	12		
Serpens	Serpentis	Ser	Serpent	12	11	
Sextans	Sextantis	Sex	Sextant	10		
Taurus	Tauri	Tau	Bull	9	3	
Telescopium	Telescopii	Tel	Telescope	18	19	
Triangulum	Trianguli	Tri	Triangle	2		
Triangulum Australe	Trianguli Australis	TrA	Southern Triangle	20		
Tucana	Tucanae	Tuc	Toucan	20		
Ursa Major	Ursae Majoris	UMa	Great Bear	4	1	
Ursa Minor	Ursae Minoris	UMi	Lesser Bear	1		
Vela	Velorum	Vel	Sail	16		
Virgo	Virginis	Vir	Virgin	11		
Volans	Volantis	Vol	Flying Fish	20		
Vulpecula	Vulpeculae	Vul	Fox	6	7	

Variable stars

The brightness of many stars varies over longer or shorter periods of time. The most common reason for this is that the size of the star actually changes; the star pulsates. A well-known type is the Cepheid, named after Delta (δ) Cephei, a yellow supergiant, regularly pulsating every few days or weeks. They are divided into two classes; the classical (Cep) and the Population II Cepheid (Cep W). The Cepheids are important to astronomers because there is a relation between their period and luminosity. The brighter a Cepheid, the longer the period. When the period is measured, we know the real luminosity (absolute brightness) of the star. By comparing this with the amount of light we actually receive (apparent brightness) we have an important tool for calculating the distance.

Another type of pulsating variable is named after the prototype Omicron (o) Ceti or Mira (M), a red giant. These variables do not have a strict period. Several other types of pulsating variables are also named after prototypes, like U Gem, R CrB, or RR Lyr. Then there is a completely different type; the eclipsing variable. Here we have a double star in mutual orbit and one component periodically moves in front, or behind the other, causing a drop in the total amount of light we receive. The best known of this type is Beta (β) Persei or Algol. Eclipsing variables are referred to in the tables as type E, divided by the shape of their light curves into E, EA, EB and EW.

A further group is termed the eruptive variables. These undergo a very sudden and very large growth in brightness. The best known are the novae and the supernovae. A nova is a very close double star, with one component being a white dwarf; a small, but very compact star. Gas from the other component flows into the white dwarf and it ignites in a huge explosion. The brightness of the star increases temporarily by thousands of times. Some novae erupt more than once. These are known as recurrent novae.

A supernova is even more spectacular. It is the catastrophic death of a very hot star. The star's life ends when it blows itself up and for a short period it shines millions of times as bright as it was. After the star has faded again the outer shells of the star form a slowly expanding nebula. The Crab nebula (M1) in Taurus (the Bull) is an example. On the atlas all variable stars with a maximum of magnitude 6.5 or brighter are plotted.

Double stars

The majority of stars are double or multiple. Sometimes two stars appear very close in the sky, but are only on the same line of sight, while their distances differ considerably. These are called optical double stars. The real physical double stars belong together and are also called binaries. They are tied together by gravity and in mutual orbit. The same goes for triple, quadruple and even larger families of stars. The apparent separation is measured in minutes and seconds of arc. One degree (°) equals 60 minutes (′) and one minute again equals 60 seconds (″). The separation in the tables (column with heading Sep) is given in seconds of arc. There is also a column with the heading PA, meaning position angle. It gives the angular position of the fainter component in relation to the brighter. The angle is measured from the north, eastward. Keep in mind that in the sky east and west are reversed. So, when north is up, east is to the left! Consequently, the position angle is measured anti-clockwise.

All double stars with integrated (combined) magnitude of 6.5 or brighter are plotted on the charts. The tables contain only a selection of the finest targets.

Open and globular clusters

Open and globular star clusters appear on the charts as yellow discs. Open clusters are shown with a dotted outline and globular clusters are shown as a solid circle with a cross through the centre. Open clusters are usually found near the plane of our milky way galaxy, so in or close to the soft blue areas on the charts, representing the brightest parts of the milky way. Open clusters are groups of young stars, often hot and bluish and their individual stars can easily be seen in a small telescope or sometimes with the naked eye. The last column of the table gives the number of stars in the cluster.

Globular clusters are quite different. They contain larger numbers of stars and are much more compact. They are found outside the galaxy plane and the stars are older. All clusters down to magnitude 10 are plotted on the charts. The diameter (Diam) in the table is given is minutes (′) of arc.

In the first column you will find the designation of the cluster. First the NGC numbers are listed (NGC = New General Catalogue) and on the charts these numbers are found without the prefix NGC. Below these come the IC numbers (IC = Index Catalogue). These have the prefix I. on the charts. Clusters from other catalogues follow on at the end of the list. Alternative namings are in the last column. The same goes for nebulae and galaxies.

Planetary and diffuse nebulae

Planetary nebulae have nothing to do with planets. The name was given because of their disc-like appearance. They are almost spherical cast-off shells of gas from very hot stars, late in their life-spans. Often the ionized gas has a greenish colour. Examples are the Ring nebula (M57) in Lyra (the Lyre) and the Helix nebula (NGC 7293) in Aquarius (the Water Carrier).

Diffuse nebulae are areas of raw materials, dust and gas, from which stars are born. These diffuse nebulae are also found along the spiral arms of the milky way, and are also visible in other nearby galaxies. Both planetary and diffuse nebulae are printed in soft green on the atlas charts.

Galaxies

The red ovals on the charts are the most remote objects; the galaxies. Galaxies are huge systems of stars, clusters and nebulae, like our own milky way galaxy. There are several types of galaxies; the elliptical (E), spiral (S), barred spiral (SB) and irregular. The E type is subdivided according to shape. E0 for the almost spherical, to E7 for the flattened lens shape. The S and SB types are subdivided according to how tight the spiral arms are wound (Fig. 3).

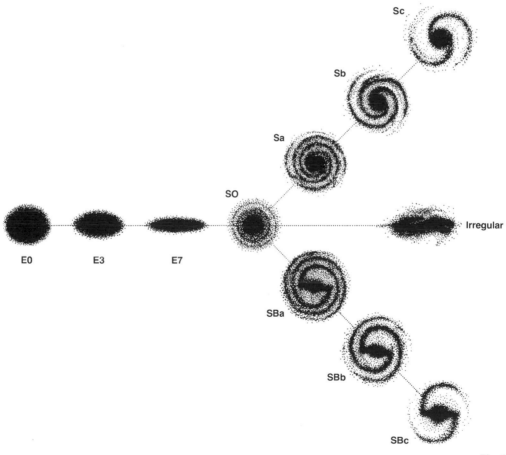

Fig. 3

The Galactic Equator

On the charts the Galactic Equator is drawn as a line of dots and dashes. It represents the projection of the plane of our galaxy on the stellar sphere. Every ten degrees of galactical longitude is marked along this equator. The centre of our milky way galaxy is at 0°.

23

The Ecliptic

The Ecliptic is the projection of the earth's orbit around the sun, or the yearly path of the sun along the sky, caused by the orbital movement of the earth. It is shown on the charts as a broken line. The moon and the planets are always close to the Ecliptic.

Abbreviations used in the star chart tables

Lists of telescopic objects are given for each map. It is clearly impossible to give a complete list, and the objects selected are those which are shown on the maps and which are within the range of telescopes of the size usually owned by amateurs. All positions are given for epoch 2000.0.

Variable stars

Range, type, period and spectrum are given.

Cep	=	classical Cepheid
CW	=	type II Cepheid
E	=	eclipsing binary
EA	=	Algol type
EB	=	Beta Lyrae type
EW	=	W Ursae Majoris type
M	=	Mira (long-period) type
SR	=	semi-regular
Irr	=	irregular
RCrB	=	R Coronae Borealis type
δ Sct	=	delta Scuti type
ZA	=	Z Andromedae type
RN	=	recurrent nova
N	=	nova
RV Tau	=	RV Tauri type

Some stars indicated as variable in the maps have visual ranges of 0.3 mag or less. These are not included in the lists.

Double stars

PA = position angle, from north (0°) through east (90°), south (180°) and west (270°). With binary stars, the values given are those for approximately the year 1990, though with binaries of reasonably short period these data will alter quite quickly

Sep = separation, in seconds of arc. The data are for the most recent available measurements

Mag = magnitudes of the components

Open clusters

NGC	=	New General Catalogue
Diam	=	diameter, in minutes of arc
Mag	=	approximate total visual magnitude
N*	=	approximate number of stars

24

Globular clusters

NGC = New General Catalogue
Diam = diameter, in minutes of arc
Mag = approximate integrated visual magnitude

Planetary nebulae

NGC = New General Catalogue
Diam = diameter, in seconds of arc
Mag = integrated photographic magnitude
Mag* = photoelectric magnitude of the central star

Nebulae

NGC = New General Catalogue
Diam = approximate maximum and minimum angular dimensions, in minutes of arc
Mag* = approximate magnitude of the illuminating star

Galaxies

NGC = New General Catalogue
Mag = approximate total integrated visual magnitude
Diam = major and minor diameters, in minutes of arc (again, bound to be somewhat arbitrary)
Type = the type according to the classical Hubble system

The lists apply to the main regions of each map. Objects such as old novae are not included, since in general they lie beyond the range of any but very powerful telescopes; also excluded are double stars of very small separation (below 0″.1) or with very faint secondary components.

The constellation abbreviations follow the official IAU practice. (See the list of constellations in Table D.)

Chart 1. Far north; declination above +70°

Variable stars

		RA h m	Dec °	Range	Type	Period d	Spectrum
VZ	Cam	07 31.1	+82 25	4.8–5.2	SR	23.7	M
YZ	Cas	00 45.7	+74 59	5.7–6.1	EA	4.47	A+F
UX	Dra	19 21.6	+76 34	5.9–7.1	SR	168	N

Double stars

		RA	Dec	PA	Sep	Mag	
48	Cas	02 02.0	+70 54	234	0.9	4.7, 6.9	Binary, 60y
49	Cas	02 05.5	+76 07	246	5.4	5.3,12.3	
β	Cep	21 28.7	+70 34	249	13.3	3.2, 7.9	
κ	Cep	20 08.9	+77 43	122	7.4	4.4, 8.4	
π	Cep	23 07.9	+75 23	346	1.2	4.6, 6.6	Slow binary
ε	Dra	19 48.2	+70 16	015	3.1	3.8, 7.4	
ψ	Dra	17 41.9	+72 09	015	30.3	4.9, 6.1	
α	UMi	02 31.8	+89 16	218	18.4	2.0, 9.0	Polaris
5	UMi	14 27.5	+75 42	{AB 124	21.7	4.3,13.3	
				{AC 131	58.8	9.8	

Open cluster

NGC	RA	Dec	Diam	Mag	N*
188 Cep	00 44.4	+85 20	14	8.1	120

Planetary nebulae

NGC	RA	Dec	Diam	Mag	Mag*
IC 3568 Cam	12 32.9	+82 33	6	11.6	12.3
40 Cep	00 13.0	+72 32	37	10.7	11.6

Galaxies

NGC	RA	Dec	Mag	Diam	Type
2146 Cam	06 18.7	+78 21	10.5	6.0×3.8	SBb
2655 Cam	08 55.6	+78 13	10.1	5.1×4.4	SBa
2715 Cam	09 08.1	+78 05	11.4	5.0×1.9	Sc
2985 UMa	09 50.4	+72 17	10.5	4.5×3.4	Sb
3147 Dra	10 16.9	+73 24	10.6	4.0×3.5	Sb
6503 Dra	17 49.4	+70 09	10.2	6.2×2.3	Sb

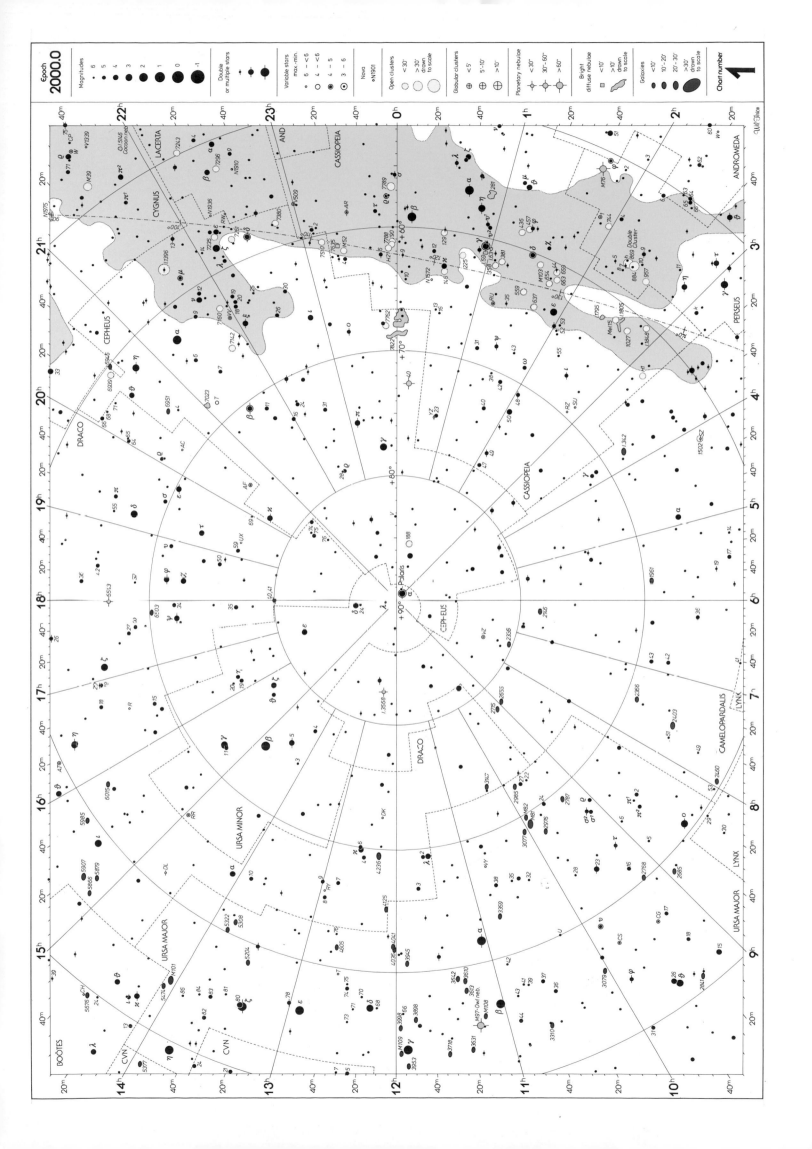

Chart 2. RA 0h to 4h. Dec +20° to +70°

Variable stars

		RA h m	Dec °	Range	Type	Period d	Spectrum
R	And	00 24.0	+38 35	5.8–14.9	M	409.3	M
W	And	02 17.6	+44 18	6.7–14.6	M	395.9	S
α	Cas	00 40.5	+56 32	?2.2– 2.5	Suspected	–	K
γ	Cas	00 56.7	+60 43	1.6– 3.3	Irr	–	B
RZ	Cas	02 48.9	+69 38	6.2– 7.7	EA	1.19	A
β	Per	03 08.2	+40 57	2.2– 3.4	EA	2.87	B+G Algol
ρ	Per	03 05.2	+38 50	3.0– 4.0	SR	50	M
X	Per	03 55.4	+31 03	6.0– 7.0	Irr	–	09.5
R	Tri	02 37.0	+34 16	5.4–12.6	M	266.5	M

Double stars

		RA	Dec	PA	Sep	Mag
γ	And	02 03.9	+42 20	063	9.8	2.3, 4.8
γ²	And			106	0.5	5.5, 6.3 Binary, 61.1 y
δ	And	00 39.3	+30 52	298	28.7	3.3,12.4
π	And	00 36.9	+33 43	173	35.9	4.4, 8.6
φ	And	01 09.5	+47 15	130	0.5	4.6, 5.5 Binary, 372 y
ω	And	01 27.7	+45 24	{AB 117 / AC 111	2.0 / 119.0	4.8,11.5 / 10.2
55	And	01 52.2	+40 44	356	59.8	5.6,10.9
ε	Ari	02 59.2	+21 20	191	1.5	5.2, 5.5 C:12.7,146"
30	Ari	02 37.0	+24 39	274	38.6	6.6, 7.4
33	Ari	02 40.7	+27 04	000	28.6	5.5, 8.4
41	Ari	02 50.0	+27 38	{AB 277 / AC 213	24.6 / 31.3	3.6,10.7 / 10.5
α	Cas	00 40.5	+56 32	275	19.8	2.3v?,13.7 Optical
γ	Cas	00 56.7	+60 43	248	2.1	2v, 11.2
η	Cas	00 49.1	+57 49	293	12.2	3.4, 7.5 Binary, 480 y
ι	Cas	02 29.1	+67 24	232	2.4	4.6, 6.9 Binary, 840 y
λ	Cas	01 31.8	+54 31	176	0.5	5.3, 5.6
ψ	Cas	01 25.9	+68 08	{AC 113 / AD 118	25.0 / 22.8	4.7, 9.6 / 9.7
35	Cas	01 21.1	+64 40	344	55.5	6.3, 8.7
γ	Per	03 04.8	+53 30	326	57.0	2.9,10.6
ε	Per	03 57.9	+40 01	010	8.8	2.9, 8.1
ζ	Per	03 54.1	+31 53	{AB 208 / AC 286 / AD 195 / AE 185	12.9 / 32.8 / 94.2 / 120.3	2.9, 9.5 / 11.3 / 9.5 / 10.2
η	Per	02 50.7	+55 54	300	28.3	3.3, 8.5
θ	Per	02 44.2	+49 14	215	19.8	4.1, 9.9 Binary, 2720 y
o	Per	03 44.3	+32 17	037	1.0	3.8, 8.3
τ	Per	02 54.3	+52 46	106	51.7	3.0,10.6
ψ	Psc	01 05.6	+21 28	159	30.0	5.6, 5.8
6	Tri	02 12.4	+30 18	071	3.9	5.3, 6.9

Open clusters

NGC		RA	Dec	Diam	Mag	N*
129	Cas	00 29.9	+60 14	21	6.5	35 Contains DL Cas
133	Cas	00 31.2	+63 22	7	9.4	5
146	Cas	00 33.1	+63 18	7	9.1	20
381	Cas	01 08.3	+61 35	6	9.3	50
436	Cas	01 15.6	+58 49	6	8.8	30
457	Cas	01 19.1	+58 20	13	6.4	80 φ Cas cluster
559	Cas	01 29.5	+63 18	4.4	9.5	60
581	Cas	01 33.2	+60 42	6	7.4	25 M103
637	Cas	01 42.9	+64 00	3.5	8.2	20
654	Cas	01 44.1	+61 53	5	6.5	60
659	Cas	01 44.2	+60 42	5	7.9	40
663	Cas	01 46.0	+61 15	16	7.1	80
744	Per	01 58.4	+55 29	11	7.9	20
752	And	01 57.8	+37 41	50	5.7	60
{869	Per	02 19.0	+57 09	30	4.3	200} Sword-handle of Perseus
{884	Per	02 22.4	+57 07	30	4.4	150}
957	Per	02 33.6	+57 32	11	7.6	30
1027	Cas	02 42.7	+61 33	20	6.7	40
1039	Per	02 42.0	+42 47	35	5.2	60 M34
1245	Per	03 14.7	+47 15	10	8.4	200
1432/5	Tau	03 47.0	+24 07	110	1.2	300+ M45 Pleiades
1444	Per	03 49.4	+52 40	4	6.6	–
IC 1805	Cas	02 32.7	+61 27	22	6.5	40

Planetary nebula

NGC		RA	Dec	Diam	Mag	
650–1	Per	01 42.4	+51 34	65×290	12.2	17 Little Dumbbell M76

Nebulae

NGC		RA	Dec	Diam	Mag*
281	Cas	00 52.8	+56 36	35×30	8
IC 1805	Cas	02 33.4	+61 26	60×60	–
IC 1848	Cas	02 51.3	+60 25	60×30	–

Galaxies

NGC		RA	Dec	Mag	Diam	Type
147	Cas	00 33.2	+48 30	9.3	12.9× 8.1	dE4
185	Cas	00 39.0	+48 20	9.2	11.5× 9.8	dE0
205	And	00 40.4	+41 41	8.0	17.4× 9.8	E6 M110 Companion to M31
221	And	00 42.7	+40 52	8.2	7.6× 5.8	E2 M32 Companion to M31
224	And	00 42.7	+41 16	3.5	178 ×63	Sb M31
598	Tri	01 33.9	+30 39	5.7	62 ×39	Sc M33 Pinwheel
891	And	02 22.6	+42 21	9.9	13.5× 2.8	Sb
925	Tri	02 27.3	+33 35	10.0	9.8× 6.0	SBc
976	Ari	02 34.0	+20 59	12.4	1.7× 1.5	Sb
1003	Per	02 39.3	+40 52	11.5	5.4× 2.1	Sc
1023	Per	02 40.0	+39 04	9.5	8.7× 3.3	E7
IC 342	Cam	03 46.8	+68 06	9.2	17.8×17.4	SBc

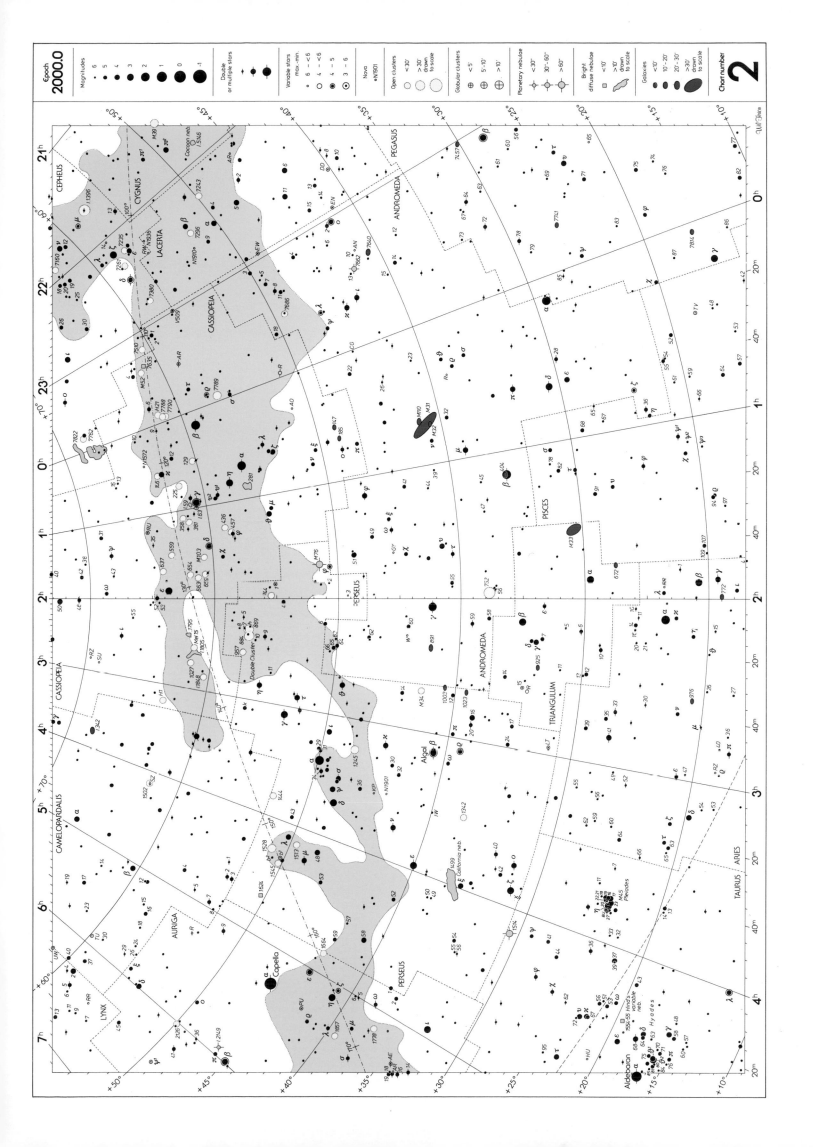

Chart 3. *RA 4h to 8h. Dec +20° to +70°*

Variable stars

		RA h m	Dec °	Range	Type	Period d	Spectrum
ε	Aur	05 02.0	+43 49	2.9– 3.8	E	9892	A–F
ζ	Aur	05 02.5	+41 05	3.7– 4.0	E	972.1	K+B
R	Aur	05 17.3	+53 35	6.7–13.9	M	457.5	M
RT	Aur	06 28.6	+30 30	5.0– 5.8	Cep	3.73	F–G
UU	Aur	06 36.5	+38 27	5.1– 6.8	SR	234	N
WW	Aur	06 32.5	+32 27	5.8– 6.5	EA	2.53	A+A
ζ	Gem	07 04.1	+20 34	3.7– 4.1	Cep	10.15	F–G
η	Gem	06 14.9	+22 30	3.2– 3.9	SR	233	M
R	Gem	07 07.4	+22 42	6.0–14.0	M	369.8	S
BU	Gem	06 12.3	+22 54	5.7– 7.5	Irr	—	M
RR	Lyn	06 26.4	+56 17	5.6– 6.0	EA	9.95	A
U	Ori	05 55.8	+20 10	4.8–12.6	M	372.4	M
HU	Tau	04 38.3	+20 41	5.9– 6.7	EA	2.06	A

Double stars

		RA	Dec	PA	Sep	Mag
δ	Aur	05 59.5	+54 17	AB 271	115.4	3.7, 9.5
				AC 067	197.1	9.5
θ	Aur	05 59.7	+37 13	AB 313	3.6	2.6, 7.1
				AC 297	50.0	10.6
ν	Aur	05 51.5	+39 09	206	54.6	4.9, 9.3
ω	Aur	04 59.3	+37 53	359	5.4	5.0, 8.0
R	Aur	05 17.3	+53 35	339	47.5	var, 8.6
14	Aur	05 15.4	+32 41	AB 352	11.1	5.1,11.1
				AC 226	14.6	5.1, 7.4
26	Aur	05 38.6	+30 30	AB 031	0.1	6.0, 6.3 Binary, 53.2 y
				AB+C 267	12.4	6.0, 8.0
1	Cam	04 32.0	+53 55	308	10.3	5.7, 6.8
α	Gem	07 34.6	+31 53	AB 088	2.5	1.9, 2.9 Binary, 420 y
				AC 164	72.5	1.6, 8.8
δ	Gem	07 20.1	+21 59	223	6.0	3.5, 8.2 Binary, 1200 y
ε	Gem	06 43.9	+25 08	094	110.3	3.0, 9.0
η	Gem	06 14.9	+22 30	266	1.4	var. 8.8 Binary, 474 y
κ	Gem	07 44.4	+24 24	240	7.1	3.6, 8.1
μ	Gem	06 22.9	+22 31	077	72.7	3.2, 9.8
ν	Gem	06 29.0	+20 13	329	112.5	4.2, 8.7
π	Gem	07 47.5	+33 25	AB 214	21.0	5.1,11.2
				AC 341	91.9	10.2
ρ	Gem	07 29.1	+31 47	AB 008	3.4	4.2,12.5
				AC 291	213.7	10.6
				CD 267	104.1	12.2
3	Gem	06 09.7	+23 07	339	0.5	5.8, 9.9
65	Gem	07 29.8	+27 55	288	12.8	5.0,13.5
70	Gem	07 38.5	+35 03	AB 191	100.4	5.6,11.6
				AC 100	162.0	10.6
				CD 242	1.6	11.6
				CE 207	17.5	14.1
4	Lyn	06 22.1	+59 22	AB 124	0.8	6.2, 7.7
				AB+C 096	26.2	12.9
				AB+D 356	100.4	11.0
12	Lyn	06 46.2	+59 27	AB 070	1.7	5.4, 6.0 Binary, 699 y
				AC 308	8.7	7.3
				AD 256	170.0	10.6
14	Lyn	06 53.1	+59 27	AB 260	0.4	5.7, 6.9 Binary, 480 y
				AB+C 122	181.0	11.0
19	Lyn	07 22.9	+55 17	AB 315	14.8	5.6, 6.5
				AD 003	214.9	8.9
				BC 287	74.2	10.9
24	Lyn	07 43.0	+58 43	320	54.7	5.0, 9.5
μ	Per	04 14.9	+48 25	349	14.8	4.1,11.6
56	Per	04 24.6	+33 58	022	4.2	5.9, 8.7
x+67	Tau	04 25.4	+22 18	173	339.0	4.2, 5.3
φ	Tau	04 20.4	+27 21	250	52.1	5.0, 8.4
χ	Tau	04 22.6	+25 38	024	19.4	5.5, 7.6
103	Tau	05 08.1	+24 16	AB 150	13.3	5.5,12.0
				AC 197	35.3	8.6
118	Tau	05 29.3	+25 09	AB 204	4.8	5.8, 6.6
				AC 099	141.3	11.6

Open clusters

NGC		RA	Dec	Diam	Mag	N*
1502	Cam	04 07.7	+62 20	8	5.7	45
1513	Per	04 10.0	+49 31	9	8.4	50
1528	Per	04 15.4	+51 14	40	6.4	40
1545	Per	04 20.9	+50 15	8	6.2	20
1664	Aur	04 51.1	+43 42	18	7.6	–
1746	Tau	05 03.6	+23 49	42	6.1	20
1778	Aur	05 08.1	+37 03	7	7.7	25
1857	Aur	05 20.2	+39 21	6	7.0	40
1893	Aur	05 22.7	+33 24	11	7.5	60
1912	Aur	05 28.7	+35 50	21	6.4	100+ M38
1960	Aur	05 36.1	+34 08	12	6.0	60 M36
2099	Aur	05 52.4	+32 33	24	5.6	150 M37
2126	Aur	06 03.0	+49 54	6	10.2	40
2129	Gem	06 01.0	+23 18	7	6.7	40
2168	Gem	06 08.9	+24 20	28	5.1	200 M35
2175	Ori	06 09.8	+20 19	18	6.8	60
2266	Gem	06 43.2	+26 58	7	9.5	50
2281	Aur	06 49.3	+41 04	15	5.4	30
2420	Gem	07 38.5	+21 34	10	8.3	100
IC 2157	Gem	06 05.0	+24 00	7	8.4	20

Planetary nebulae

NGC		RA	Dec	Diam	Mag	Mag*
1514	Tau	04 09.2	+30 47	114	10	9.4
2392	Gem	07 29.2	+20 55	13×44	10	10.5 Eskimo Nebula

Nebulae

NGC		RA	Dec	Diam	Mag*	
1499	Per	04 00.7	+36 37	145×40	4	California Nebula
1952	Tau	05 34.5	+22 01	6× 4	16	M1 Crab Nebula: SNR
IC 405	Aur	05 16.2	+34 16	30×19	6v	AE Aur: Flaming Star Nebula

Galaxies

NGC		RA	Dec	Mag	Diam	Type
1961	Cam	05 42.1	+69 23	11.1	4.3× 3.0	Sbp
2366	Cam	07 28.9	+69 13	10.9	7.6× 3.5	Irr
2403	Cam	07 36.9	+65 36	8.4	17.8×11.0	Sc
2460	Cam	07 56.9	+60 21	11.7	2.9× 2.2	Sb

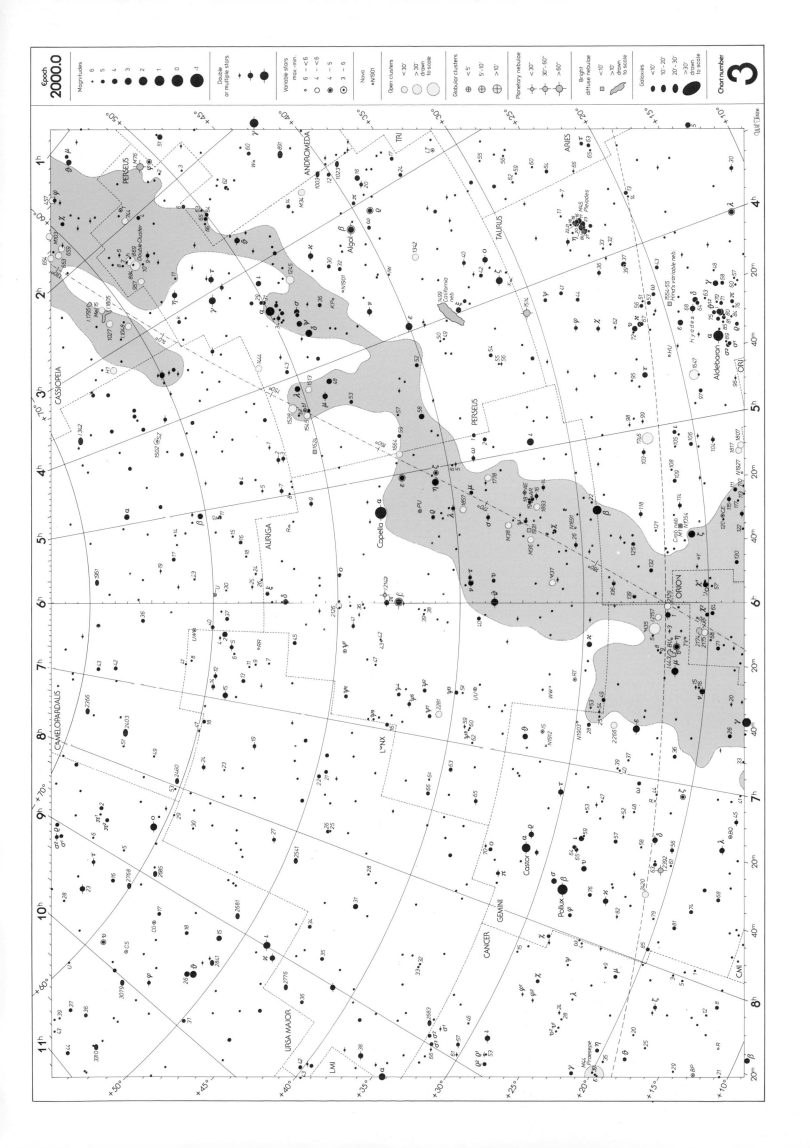

Chart 4. RA 8h to 12h. Dec +20° to +70°

Variable stars

	RA h m	Dec °	Range	Type	Period d	Spectrum
RS Cnc	09 10.6	+30 58	6.2- 7.7	SR	120	M
UY Leo	10 29.4	+23 04	9.5-11.0	Irr	—	M
R LMi	09 45.6	+34 31	6.3-13.2	M	371.9	M
ST UMa	11 27.8	+45 11	7.7- 9.5	SR	81	M
VY UMa	10 45.7	+67 25	5.9- 6.5	Irr	—	N

Double stars

	RA	Dec	PA	Sep	Mag
β LMi	10 27.9	+36 42	250	0.2	4.4, 6.1 Binary, 37.2 y
11 LMi	10 35.7	+35 49	031	2.0	5.4,13.9
40 LMi	10 43.0	+26 20	112	18.4	5.5,12.5
38 Lyn	09 18.8	+36 48	AB 229	2.7	3.9, 6.6
			BC 212	87.7	10.8
			BD 256	177.9	10.7
α UMa	11 03.7	+61 45	283	0.7	1.9, 4.8 Binary, 44.7 y
ι UMa	08 59.2	+48 02	100	1.8	3.1,10.2 Binary, 818 y
ϰ UMa	09 03.6	+47 09	258	0.1	4.2, 4.4 Binary, 70 y
o UMa	08 30.3	+60 43	192	7.1	3.4,15
ν UMa	11 18.5	+33 06	147	7.2	3.5, 9.9
ξ UMa	11 18.2	+31 32	060	1.3	4.3, 4.8 Binary, 59.8 y
φ UMa	09 52.1	+54 04	188	0.2	5.3, 5.4 Binary, 105.5 y
23 UMa	09 31.5	+63 04	AB 270	22.7	3.7, 8.9
			AC 231	99.6	10.4
57 UMa	11 29.1	+39 20	AB 359	5.4	5.3, 8.3
			AC 009	216.5	11.5

Planetary nebula

NGC	RA	Dec	Diam	Mag	Mag*
3587	11 14.8	+55 01	194	12.0	15.9 M97 Owl Nebula

Galaxies

NGC	RA	Dec	Mag	Diam	Type	
2541 Lyn	08 14.7	+49 04	11.7	6.6× 3.5	S	
2681 UMa	08 53.5	+51 19	10.3	3.8× 3.5	Sa	
2683 Lyn	08 52.7	+33 25	9.7	9.3× 2.5	Sb	
2685 UMa	08 55.6	+58 44	11.0	5.2× 3.0	Sbp	
2768 UMa	09 11.6	+60 02	10.0	6.3× 2.8	E5	
2776 Lyn	09 12.2	+44 57	11.6	2.9× 2.7	Sc	
2787 UMa	09 19.3	+69 12	10.8	3.4× 2.3	Sap	
2841 UMa	09 22.0	+50 58	9.3	8.1× 3.8	Sb	
2976 LMi	09 47.3	+67 55	10.1	4.9× 2.5	Scp	
3003 LMi	09 48.6	+33 25	11.7	5.9× 1.7	SBc	
3031 UMa	09 55.6	+69 04	6.9	25.7×14.1	SB	M81
3034 UMa	09 55.8	+69 41	8.4	11.2× 4.6	Pec	M82
3077 UMa	10 03.3	+68 44	9.8	4.6× 3.6	E2p	
3079 UMa	10 02.0	+55 41	10.6	7.6× 1.7	Sb	
3184 UMa	10 18.3	+41 25	9.7	6.9× 6.8	Sc	
3190 Leo	10 18.1	+21 50	11.0	4.6× 1.8	Sb	
3198 UMa	10 19.9	+45 33	10.4	8.3× 3.7	Sc	
3245 LMi	10 27.3	+28 30	10.8	3.2× 1.9	E5	
3254 LMi	10 29.3	+29 30	11.5	5.1× 1.9	Sb	
3294 LMi	10 36.3	+37 20	11.7	3.3× 1.8	Sc	
3310 UMa	10 38.7	+53 30	10.9	3.6× 3.0	SBc	
3344 LMi	10 43.5	+24 55	9.9	6.9× 6.5	Sc	
3359 UMa	10 46.6	+63 13	10.4	6.8× 4.3	SBc	
3414 LMi	10 51.3	+27 59	10.7	3.6× 2.7	SBa	
3430 LMi	10 52.2	+32 57	11.5	3.9× 2.3	Sc	
3432 LMi	10 52.5	+36 37	11.2	6.2× 1.5	SB	
3486 LMi	11 00.4	+28 58	10.3	6.9× 5.4	Sc	
3610 UMa	11 18.4	+58 47	10.7	3.2× 2.5	E2p	
3631 UMa	11 21.0	+53 10	10.4	4.6× 4.1	Sc	
3646 Leo	11 21.7	+20 10	11.2	3.9× 2.6	Sc	
3675 UMa	11 26.1	+43 35	10.9	5.9× 3.2	Sb	
3687 UMa	11 28.0	+29 31	12.6	2.0× 2.0	SBap	
3718 UMa	11 32.6	+53 04	10.5	8.7× 4.5	SBa	
3726 UMa	11 33.3	+47 02	10.4	6.0× 4.5	Sc	
3877 UMa	11 46.1	+47 30	11.6	5.4× 1.5	Sb	
3898 UMa	11 49.2	+56 05	10.8	4.4× 2.6	Sb	
3945 UMa	11 53.2	+60 41	10.6	5.5× 3.6	SBa	
3949 UMa	11 53.7	+47 52	11.0	3.0× 1.8	Sb	
3953 UMa	11 53.8	+52 20	10.1	6.6× 3.6	Sb	
3998 UMa	11 57.9	+55 27	10.6	3.1× 2.5	E2p	
4026 UMa	11 59.4	+50 58	11.7	5.1× 1.4	SO	

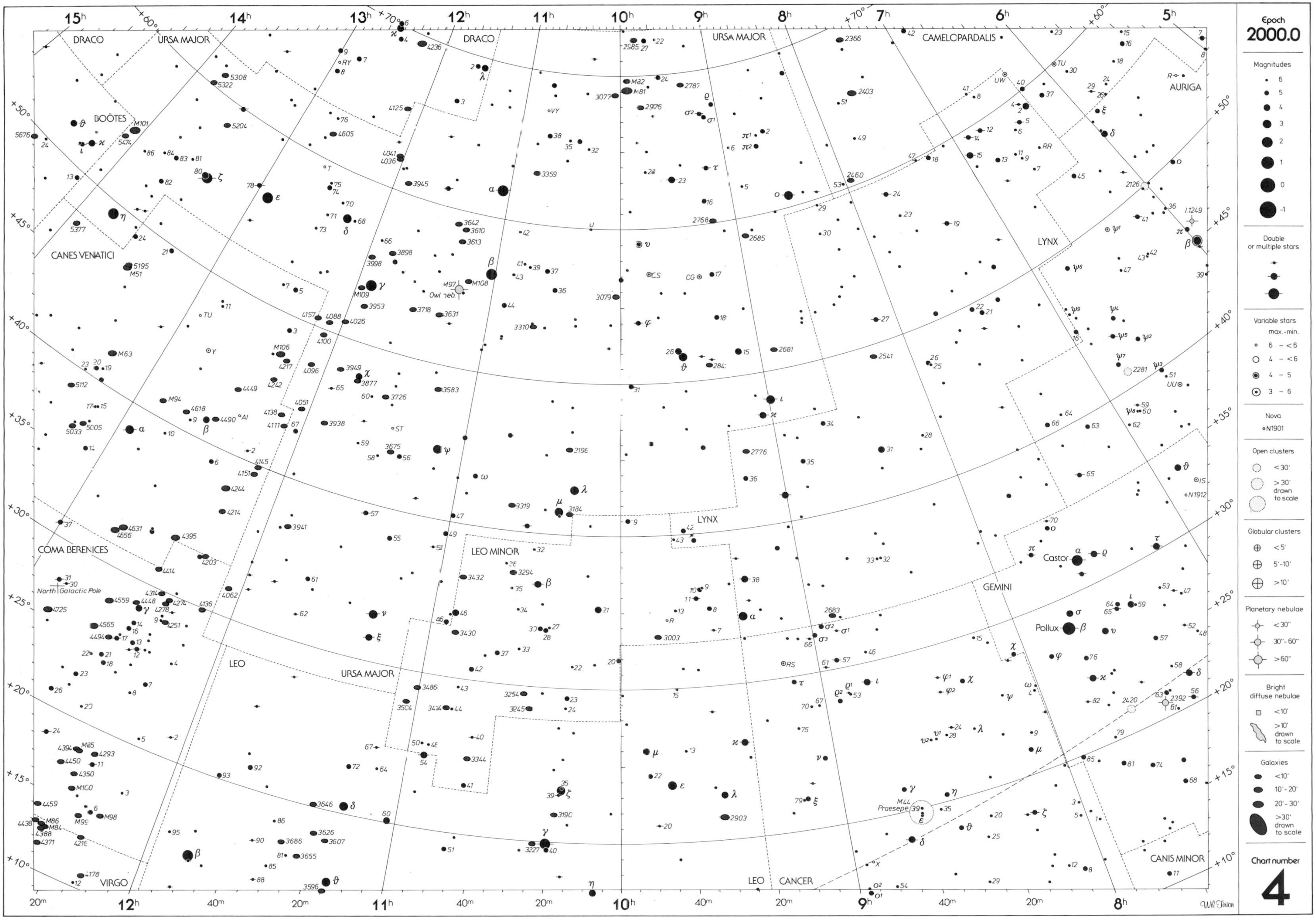
Epoch
2000.0
Magnitudes
Double or multiple stars
Variable stars max.-min.
Nova
Open clusters
Globular clusters
Planetary nebulae
Bright diffuse nebulae
Galaxies
Chart number
4
DRACO
URSA MAJOR
BOÖTES
CANES VENATICI
COMA BERENICES
LEO
LEO MINOR
LYNX
CAMELOPARDALIS
AURIGA
GEMINI
CANCER
CANIS MINOR
VIRGO
Castor
Pollux
Praesepe
North Galactic Pole
Owl neb.
Wil Tirion

Chart 5. RA 12h to 16h. Dec +20° to +70°

Variable stars

Name	RA h m	Dec °	Range	Type	Period d	Spectrum
W Boö	14 43.4	+26 32	4.7– 5.4	SR	450	M
i(44) Boö	15 03.8	+47 39	6.5– 7.1	EW	0.27	G+G
R CVn	13 49.0	+39 33	6.5–12.9	M	328.5	M
Y CVn	12 45.1	+45 26	7.4–10.0	SR	157	N La Superba
TU CVn	12 54.9	+47 12	5.6– 6.6	SR	50	M
FS Com	13 06.4	+22 37	5.3– 6.1	SR	58	M (FZ in map)
R CrB	15 48.6	+28 09	5.7–15	RCrB	–	F8p
T CrB	15 59.5	+25 55	2.0–10.8	RN	29 000?	M+P(Q) Blaze Star
RY Dra	12 56.4	+66 00	5.6– 8.0	SR	173	N
T UMa	12 36.4	+59 29	6.6–13.4	M	256.5	M
RR UMi	14 57.6	+65 56	6.1– 6.5	SR?	40	M

Double stars

Name	RA	Dec	PA	Sep	Mag
γ Boö	14 32.1	+38 19	111	33.4	3.0,12.7
ε Boö	14 45.0	+27 04	339	2.8	2.5, 4.9
κ Boö	14 13.5	+51 47	236	13.4	4.6, 6.6
μ Boö	15 24.5	+37 23	171	108.3	4.3, 7.0
39 Boö	14 49.7	+48 43	045	2.9	6.2, 6.9
i(44) Boö	15 03.8	+47 39	040	1.0	5.3v,6.2 Binary, 225 y
α CVn	12 56.0	+38 19	229	19.4	2.9, 5.5 Cor Caroli
2 CrB	12 16.1	+40 40	260	11.4	5.8, 8.1
ε CrB	15 57.6	+26 53	AB 003 / AC 174	1.8 / 101.4	4.2,12.6 / 11.5
η CrB	15 23.2	+30 17	AB 030 / AC 012 / AB+D 047	1.0 / 57.7 / 215.0	5.6, 5.9 Binary, 41.6 y / 12.5 / 10.9
ζ UMa	13 23.9	+54 56	AB 152 / AC 071	14.4 / 708.7	2.3, 4.0 Mizar / 2.1, 4.0 Mizar/Alcor
78 UMa	13 00.7	+56 22	057	1.5	5.0, 7.4 Binary, 116 y

Globular cluster

NGC	RA	Dec	Diam	Mag	Type
5272 CVn	13 42.2	+28 23	16.2	6.4	M3

Galaxies

NGC	RA	Dec	Mag	Diam	Type
4036 UMa	12 01.4	+61 54	10.6	4.5× 2.0	E6
4041 UMa	12 02.2	+62 08	11.1	2.8× 2.7	Sc
4051 UMa	12 03.2	+44 32	10.3	5.0× 4.0	Sc
4062 UMa	12 04.1	+31 54	11.2	4.3× 2.0	Sb
4088 UMa	12 05.6	+50 33	10.5	5.8× 2.5	Sc
4096 UMa	12 06.0	+47 29	10.6	6.5× 2.0	Sc
4100 UMa	12 06.2	+49 35	11.5	5.2× 1.9	Sb
4111 CVn	12 07.1	+43 04	10.8	4.8× 1.1	S0
4125 Dra	12 08.1	+65 11	9.8	5.1× 3.2	E5p
4136 Com	12 09.3	+29 56	11.7	4.1× 3.9	Sc
4138 Com	12 09.5	+43 41	12.3	2.9× 1.9	E4
4145 CVn	12 10.0	+39 53	11.0	5.8× 4.4	Sc
4151 CVn	12 10.5	+39 24	10.4	5.9× 4.4	Pec
4214 CVn	12 15.6	+36 20	9.8	7.9× 6.3	Irr
4217 CVn	12 15.8	+47 06	11.9	5.5× 1.8	Sb
4236 Dra	12 16.7	+69 28	9.7	18.6× 6.9	Sb
4242 CVn	12 17.5	+45 37	11.0	4.8× 3.8	S
4244 CVn	12 17.5	+37 49	10.2	16.2× 2.5	S
4251 Com	12 18.1	+28 10	11.6	4.2× 1.9	E7
4258 CVn	12 19.0	+47 18	8.3	18.2× 7.9	Sb M106
4278 Com	12 20.1	+29 17	10.2	3.6× 3.5	E1
4314 Com	12 22.6	+29 53	10.5	4.8× 4.3	SBa
4395 CVn	12 25.8	+33 33	10.1	12.9×11.0	S
4448 Com	12 28.2	+28 37	11.1	4.0× 1.6	Sb
4449 CVn	12 28.1	+44 06	9.4	5.1× 3.7	Irr
4490 CVn	12 30.6	+41 38	9.8	5.9× 3.1	Sc
4494 Com	12 31.4	+25 47	9.9	4.8× 3.8	Sb
4559 Com	12 36.0	+27 58	9.8	10.5× 4.9	Sc
4565 Com	12 36.3	+25 59	9.6	16.2× 2.8	Sb
4605 UMa	12 40.0	+61 37	11.0	5.5× 2.3	SBcp
4618 CVn	12 41.5	+41 09	10.8	4.4× 3.8	Sc
4631 CVn	12 42.1	+32 32	9.3	15.1× 3.3	Sc
4656–7 CVn	12 44.0	+32 10	10.4	13.8× 3.3	SBp
4725 Com	12 50.4	+25 30	9.2	11.0× 7.9	Sb
4736 CVn	12 50.9	+41 07	8.2	11.0× 9.1	Sb M94
4826 Com	12 56.7	+21 41	8.5	9.3× 5.4	Sab M64 Black-Eye
5005 CVn	13 10.9	+37 03	9.8	5.4× 2.7	Sb
5033 CVn	13 13.4	+36 36	10.1	10.5× 5.6	Sb
5055 CVn	13 15.8	+42 02	8.6	12.3× 7.6	Sb M63
5112 CVn	13 21.9	+38 44	11.9	3.9× 2.9	Sc
5194 CVn	13 29.9	+47 12	8.4	11.0× 7.8	Sc M51 Whirlpool
5195 CVn	13 30.0	+47 16	9.3	5.4× 4.2	Pec Companion of M51
5308 UMa	13 47.0	+60 58	11.3	3.5× 0.8	S0
5322 UMa	13 49.3	+60 12	10.0	5.5× 3.9	E2
5371 CVn	13 55.7	+40 28	10.7	4.4× 3.6	Sb
5457 UMa	14 03.2	+54 21	7.7	26.9×26.3	Sb M101 Pinwheel
5475 UMa	14 05.2	+55 45	13.4 photo.	2.2× 0.6	Sa
5676 Boö	14 32.8	+49 28	10.9	3.9× 2.0	Sc
5866 Dra	15 06.5	+55 46	10.0	5.2× 2.3	E6p
5879 Dra	15 09.8	+57 00	11.5	4.4× 1.7	Sb
5907 Dra	15 15.9	+56 19	10.4	12.3× 1.8	Sb
5985 Dra	15 39.6	+59 20	11.0	5.5× 3.2	Sb
6015 Dra	15 51.4	+62 19	11.2	5.4× 2.3	Sc

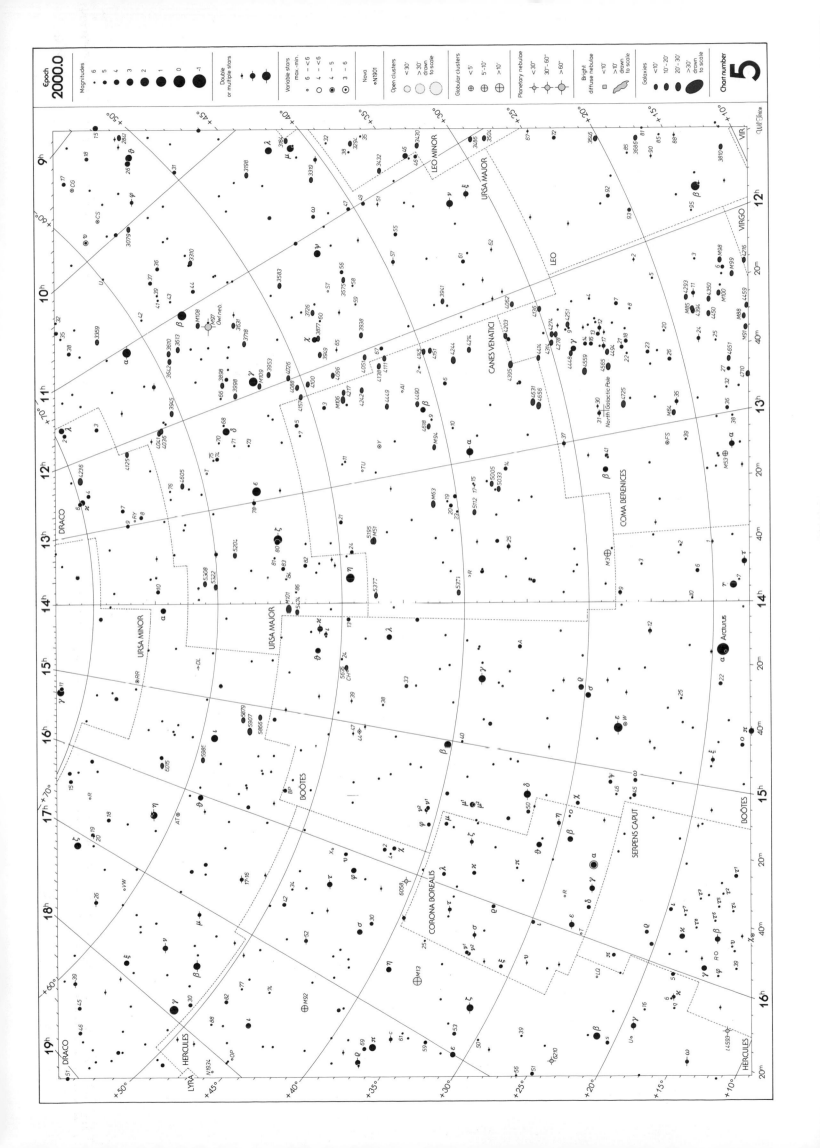

Chart 6. RA 16h to 20h. Dec +20° to +70°

Variable stars

	RA h m	Dec °	Range	Type	Period d	Spectrum
χ Cyg	19 50.6	+32 55	3.3–14.2	M	406.9	S
R Cyg	19 36.8	+50 12	6.1–14.2	M	426.4	M
AF Cyg	19 30.2	+46 09	7.4– 9.4	SR	94	M
RT Cyg	19 43.6	+48 47	6.4–12.7	M	190.2	M
R Dra	16 32.7	+66 45	6.7–13.0	M	245.5	M
AT Dra	16 17.3	+59 45	6.8– 7.5	Irr	—	M
VW Dra	17 16.5	+60 40	6.0– 6.5	SR	170	K
30 (g) Her	16 28.6	+41 53	5.7– 7.2	SR	70	M
68 (u) Her	17 17.3	+35 06	4.6– 5.3	EB	2.05	B+B
X Her	16 02.7	+47 14	7.5– 8.6	SR	95	M
β Lyr	18 50.1	+33 22	3.3– 4.3	EB	12.94	B+A
R Lyr	18 55.3	+43 57	3.9– 5.0	SR	46	M
XY Lyr	18 38.1	+39 40	7.3– 7.8	Irr	—	M
U Vul	19 36.6	+20 20	6.8– 7.5	Cep	7.99	F–G

Double stars

	RA	Dec	PA	Sep	Mag
β Cyg	19 30.7	+27 58	054	34.4	3.1, 5.1 Albireo
δ Cyg	19 45.0	+10 46	225	2.4	2.9, 6.3 Binary, 828 y
η Cyg	19 56.3	+35 05		7.4	3.9,11.9
			AB 208	46.0	10.4
			AC 327	49.7	10.4
			AD 169	60.2	11.4
			AE 246		
ψ Cyg	19 55.6	+52 26	178	3.2	4.9, 7.4
17 Cyg	19 46.4	+33 44	069	26.0	5.0, 9.2
β Dra	17 30.4	+52 18	AB 013	4.2	2.8,13.8
			AC 156	117.4	12.5
η Dra	16 24.0	+61 31	142	5.2	2.7, 8.7
μ Dra	17 05.3	+54 28	020	1.9	5.7, 5.7 Binary, 482 y
ν Dra	17 32.2	+55 11	312	61.9	4.9, 4.9
ο Dra	18 51.2	+59 23	326	34.2	4.8, 7.8
ω Dra	17 36.9	+68 45	276	72.3	4.9,13.2
17 Dra	16 36.2	+52 55	AB 108	3.4	5.4, 6.4 } C is 16 Dra
			AC 194	90.3	5.5 }
39 Dra	18 23.9	+58 48	AB 351	3.8	5.0, 8.0
			AC 021	88.9	7.4
			AE 066	198.0	10.9
			AF 080	150.3	11.2
			AG 254	36.0	14.2
			AH 086	41.3	14.2
δ Her	17 15.0	+24 50	236	8.9	3.7, 8.2 Optical
ζ Her	16 41.3	+31 36	089	1.6	2.9, 5.5 Binary, 34.5 y
μ Her	17 46.5	+27 43	247	33.8	3.4,10.1
ρ Her	17 23.7	+37 09	316	4.1	4.6, 5.6
τ Her	16 19.7	+46 19	146	6.7	3.9,14.6
68 (u) Her	17 17.3	+33 06	060	4.4	4.8v,10.2
90 Her	17 53.3	+40 00	116	1.6	5.2, 8.5
95 Her	18 01.5	+21 36	258	6.3	5.0, 5.1
100 Her	18 07.8	+26 06	183	14.2	5.9, 6.0
102 Her	18 08.8	+20 49	136	23.4	4.4,11.9
δ¹ Lyr	18 53.7	+36 58	020	174.6	5.6, 9.3
δ² Lyr	18 54.5	+36 54	349	86.2	4.5,11.2
ε Lyr	18 44.3	+39 40	AB+CD 173	207.7	4.7, 5.1
			AB 357	2.6	5.0, 6.1
			CD 094	2.3	5.2, 5.5
ζ Lyr	18 44.8	+37 36	150	43.7	4.3, 5.9
η Lyr	19 13.8	+39 09	082	28.1	4.4, 9.1
α–8 Vul	19 28.7	+24 40	028	413.7	4.4, 5.8
2 (ES) Vul	19 17.7	+23 02	127	1.8	5.4, 9.2

Open clusters

NGC	RA	Dec	Diam	Mag	N*
6791 Lyr	19 20.7	+37 51	16	9.5	300
6811 Cyg	19 38.2	+46 34	13	6.8	70
6819 Cyg	19 41.3	+40 11	5	7.3	—
6823 Vul	19 43.1	+23 18	12	7.1	30
6830 Vul	19 51.0	+23 04	12	7.9	20
6834 Cyg	19 52.2	+29 25	5	7.8	50

Globular clusters

NGC	RA	Dec	Diam	Mag	
6205 Her	16 41.7	+36 28	16.6	5.9	M13
6341 Her	17 17.1	+43 08	11.2	6.5	M92
6779 Lyr	19 16.6	+30 11	7.1	8.2	M56

Planetary nebulae

NGC	RA	Dec	Diam	Mag	Mag*	
6058 Her	16 04.4	+40 41	23	13.3	13.8	
6210 Her	16 44.5	+23 49	14	9.3	12.9	
6543 Dra	17 58.6	+66 38	18×350	8.8	11.4	
6720 Lyr	18 53.6	+33 02	70×150	9.7	14.8	M57 Ring Nebula
6826 Cyg	19 44.8	+50 31	30×140	9.8	10.4	Blinking Nebula
6853 Vul	19 59.6	+22 43	350×910	7.6	13.9	M27 Dumbbell Nebula

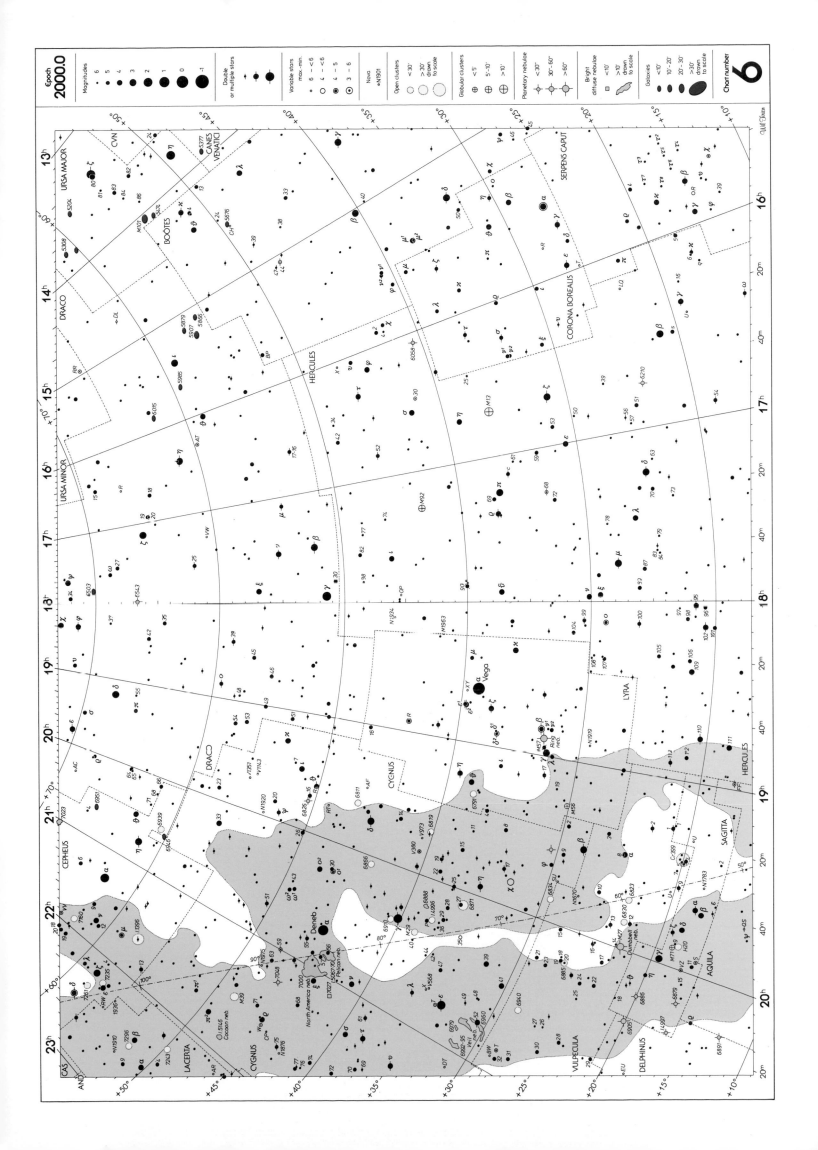

Chart 7. RA 20h to 0h. Dec +20° to +70°

Variable stars

		RA h m	Dec °	Range	Type	Period d	Spectrum
ρ	Cas	23 53.4	+58 30	4.1– 6.2	?	—	F–K
R	Cas	23 58.4	+51 24	4.7–13.5	M	430.5	M
δ	Cep	22 29.2	+58 25	3.5– 4.4	Cep	5.37	F–G
T	Cep	21 09.5	+68 29	5.2–11.3	M	388.1	M
RW	Cep	22 23.1	+55 58	8.6–10.7	SR	346	K
VV	Cep	21 56.7	+63 38	4.8– 5.4	E	7430	M+B
μ	Cep	21 43.5	+58 47	3.4– 5.1	SR?	730	M
P	Cyg	20 17.8	+38 02	3.0– 6.0	Irr	—	Bp
T	Cyg	20 47.2	+34 22	5.0– 5.5	Irr	—	K
X	Cyg	20 43.4	+35 35	5.9– 6.9	Cep	16.39	F–G
β	Peg	23 03.8	+28 05	2.3– 2.7	SR	38	M

Double stars

		RA	Dec	PA	Sep	Mag
κ	And	23 40.4	+44 20	AB 194	46.8	4.1,11.1
				AC 294	107.0	11.1
2	And	23 02.6	+42 45	AB 345	0.4	5.1, 8.8 Binary, 77 y
				AC 192	90.4	13.7
σ	Cas	23 59.0	+55 45	326	3.0	5.0, 7.1
6	Cas	23 48.8	+62 13	193	1.6	5.5, 8.0
δ	Cep	22 29.2	+58 25	191	41.0	var, 7.5 Optical
η	Cep	20 45.3	+61 50	066	51.7	3.4,11.1 Optical
ξ	Cep	22 03.8	+64 38	277	7.7	4.4, 6.5 Binary, 3800 y
o	Cep	23 18.6	+68 07	220	2.9	4.9, 7.1 Binary, 796 y
ε	Cyg	20 46.2	+33 58	272	54.9	2.5,11.5
τ	Cyg	21 14.8	+38 03	015	0.5	3.8, 6.4 Binary, 50 y
μ	Cyg	21 44.1	+28 45	300	1.6	4.8, 6.1 Binary, 507 y
ν	Cyg	21 17.9	+34 54	AB 220	15.1	4.4,10.0
				AC 181	21.5	10.0
69	Cyg	21 25.8	+36 40	AB 030	33.0	5.9,10.3
				AC 098	54.0	9.0
8	Lac	22 35.9	+39 38	AB 186	22.4	5.7, 6.5
				AC 169	48.8	10.5
				AD 144	81.8	9.3
				AE 239	336.8	7.8
κ	Peg	21 44.6	+25 39	095	0.3	4.7, 5.0 Binary, 11.6 y
2	Peg	21 29.9	+23 38	332	29.8	4.6,11.6
32	Peg	22 21.3	+28 20	AB 127	72.6	4.8, 9.1
				AD 307	42.3	11.8
				AE 116	60.3	11.8
				BC 018	2.4	10.8
72	Peg	23 34.0	+31 20	085	0.5	5.7, 5.8 Binary, 241 y
78	Peg	23 44.0	+29 22	235	1.0	5.0, 8.1
16	Vul	20 02.0	+24 56	115	0.8	5.8, 6.2

Open clusters

NGC	RA	Dec	Diam	Mag	N*	
6871 Cyg	20 05.9	+35 47	20	5.2	15	
6866 Cyg	20 03.7	+44 10	7	7.6	80	
6885 Vul	20 12.0	+26 29	7	5.7	30	
6910 Cyg	20 23.1	+40 47	8	7.4	50	
6913 Cyg	20 23.9	+38 32	7	6.6	50	M29
6939 Cyg	20 31.4	+60 38	8	7.8	80	
6940 Vul	20 34.6	+28 18	31	6.3	60	
7067 Cyg	21 24.2	+48 01	3	9.7	20	
7092 Cyg	21 32.2	+48 26	32	4.6	30	M39
7160 Cep	21 53.7	+62 36	7	6.1	12	
7235 Cep	22 12.6	+57 17	4	7.7	30	
7243 Lac	22 15.3	+49 53	21	6.4	40	
7261 Cep	22 20.4	+58 05	6	8.4	30	
7296 Lac	22 28.2	+52 17	4	9.7	20	
7510 Cep	23 11.5	+60 34	4	7.9	60	
7654 Cas	23 24.2	+61 35	13	6.9	100	M52
7686 And	23 30.2	+49 08	15	5.6	20	
7788 Cas	23 56.7	+61 24	9	9.4	20	
7789 Cas	23 57.0	+56 44	16	6.7	300	
7790 Cas	23 58.4	+61 13	17	8.5	40	
IC 1396 Cep	21 39.1	+57 30	50	3.5	50	
H.21 Cas	23 54.1	+61 46	4	9.0	6	

Globular cluster

NGC	RA	Dec	Diam	Mag	
7078 Peg	21 30.0	+12 10	12.3	6.3	M15

Planetary nebulae

NGC	RA	Dec	Diam	Mag	Mag*
7048 Cyg	21 14.2	+46 16	61	11.3	18
7662 And	23 25.9	+42 33	20	9.2	13.2

Nebulae

NGC	RA	Dec	Diam	Mag*	
6960 Cyg	20 45.7	+30 43	70× 6	—	Filamentary Nebula: 52 Cyg
6992/5 Cyg	20 56.4	+31 43	60× 8	—	Veil Nebula: SNR
7000 Cyg	20 58.8	+44 20	120×100	6	North America Nebula
7023 Cep	21 01.8	+68 12	18× 18	6.8	
7635 Cas	23 20.7	+61 12	15× 8	7	Bubble Nebula
IC 5067/70 Cyg	20 50.8	+44 21	80× 70	—	Pelican Nebula
IC 5146 Cyg	21 53.5	+47 16	12× 12	10	Cocoon Nebula, with sparse cluster

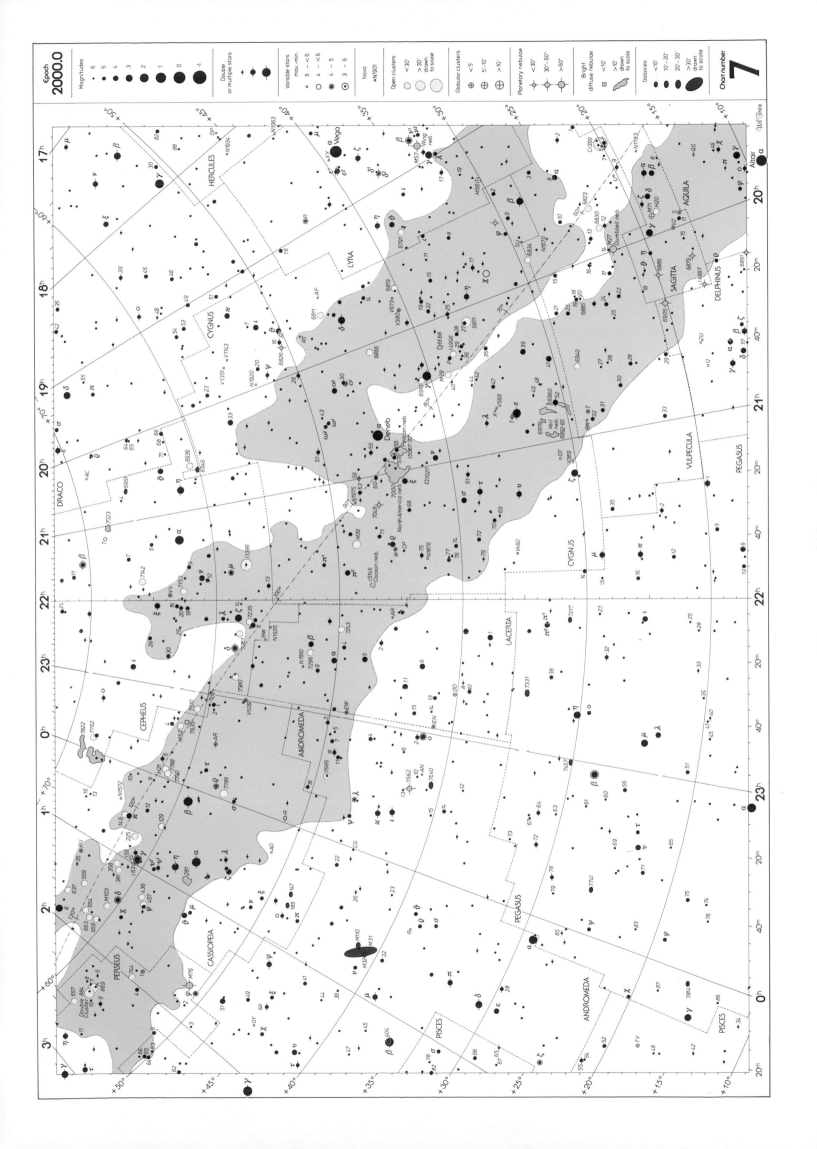

Chart 8. RA 0h to 4h. Dec +20° to −20°

Variable stars

	RA h m	Dec °	Range	Type	Period d	Spectrum
U Ari	03 11.0	+14 48	7.2–15.2	M	371.1	M
RZ Ari	02 55.8	+18 20	5.6– 6.1	SR	30	M
o Cet	02 19.3	−02 59	1.7–10.1	M	332.0	M Mira
Z Eri	02 47.9	−12 28	7.0– 8.6	SR	80	M
TV Psc	00 28.0	+17 54	4.6– 5.4	SR	70	M

Double stars

	RA	Dec	PA	Sep	Mag	
γ Ari	01 53.5	+19 18	000	7.8	4.8, 4.8	
π Ari	02 49.3	+17 28	{AB 120	3.2	5.2, 8.7	
			AC 110	25.2	10.8	
γ Cet	02 43.3	+03 14	294	2.8	3.5, 7.3	
ε Cet	02 39.6	−11 52	039	0.1	5.8, 5.8	Binary, 2.7 y
θ Cet	01 24.0	−08 11	057	65.4	3.6,14.6	
ν Cet	02 35.9	+05 36	081	8.1	4.9, 9.5	
o Cet	02 19.3	−02 59	085	0.3	var,12.0	Binary, 400 y
χ Cet	01 49.6	−10 41	250	183.8	4.9, 6.9	
12 Cet	00 30.0	−03 57	194	10.3	5.7,10.5	
13 Cet	00 35.2	−03 36	224	0.2	5.6, 6.3	Binary, 6.9 y
26 Cet	01 03.8	+01 22	253	16.0	6.2, 8.6	
37 Cet	01 14.4	−07 55	331	49.7	5.2, 8.7	
66 Cet	02 12.8	−02 24	{AB 234	16.5	5.7, 7.5	
			AC 061	172.7	11.4	
84 Cet	02 41.2	−00 42	310	4.0	5.8, 9.0	
95 Cet	03 18.4	−00 56	245	1.2	5.6, 7.5	Binary, 217 y
ρ² Eri	03 02.7	−07 41	075	1.8	5.3, 9.5	
γ Peg	00 13.2	+15 11	285	63.4	2.8,11.7	
α Psc	02 02.0	+02 46	279	1.9	4.2, 5.1	Binary, 933 y
ζ Psc	01 13.7	+07 35	063	23.0	5.6, 6.5	
η Psc	01 31.5	+15 21	036	1.0	3.6,10.6	
ψ¹ Psc	01 05.6	+21 28	{AB 159	30.0	5.6, 5.8	
			AC 123	92.6	11.2	
51 Psc	00 32.4	+06 57	083	27.5	5.7, 9.5	

Planetary nebula

NGC	RA	Dec	Diam	Mag	Mag*
246 Cet	00 47.0	−11 53	225	8	11.9

Galaxies

NGC	RA	Dec	Mag	Diam	Type	
428 Cet	01 12.9	+00 59	11.3	4.1×3.2	Scp	
470 Psc	01 19.7	+03 25	11.9	3.0×2.0	Sc	
474 Psc	01 20.1	+03 25	11.1	7.9×7.2	SO	
488 Psc	01 21.8	+05 15	10.3	5.2×4.1	Sb	
524 Psc	01 24.8	+09 32	10.6	3.2×3.2	E1	
584 Cet	01 31.3	−06 52	10.3	3.8×2.4	E4	
628 Psc	01 36.7	+15 47	9.2	10.2×9.5	Sc	M74
720 Cet	01 53.0	−13 44	10.2	4.4×2.8	E3	
772 Ari	01 59.3	+19 01	10.3	7.1×4.5	Sb	Arp 78
864 Cet	02 15.5	+06 00	11.0	4.6×3.5	Sc	
895 Cet	02 21.6	−05 31	11.8	3.6×2.8	Sb	
936 Cet	02 27.6	−01 09	10.1	5.2×4.4	SBa	
1042 Cet	02 40.4	−08 26	10.9	4.7×3.9	Sc	
1055 Cet	02 41.8	+00 26	10.6	7.6×3.0	Sb	
1073 Cet	02 43.7	+01 23	11.0	4.9×4.6	SBc	
1084 Eri	02 46.0	−07 35	10.6	2.9×1.5	Sc	
1087 Cet	02 46.4	−00 30	11.0	3.5×2.3	Sc	
1179 Eri	03 02.6	−18 54	11.8	4.6×3.9	Sp	
1300 Eri	03 19.7	−19 25	10.4	6.5×4.3	SBp	
1337 Eri	03 28.1	−08 23	11.7	6.8×2.0	S	
1407 Eri	03 40.2	−18 35	9.8	2.5×2.5	EO	
1068 Cet	02 42.7	−00 01	8.8	6.9×5.9	SBp	M77
7814 Peg	00 03.3	+16 09	10.5	6.3×2.6	Sb	

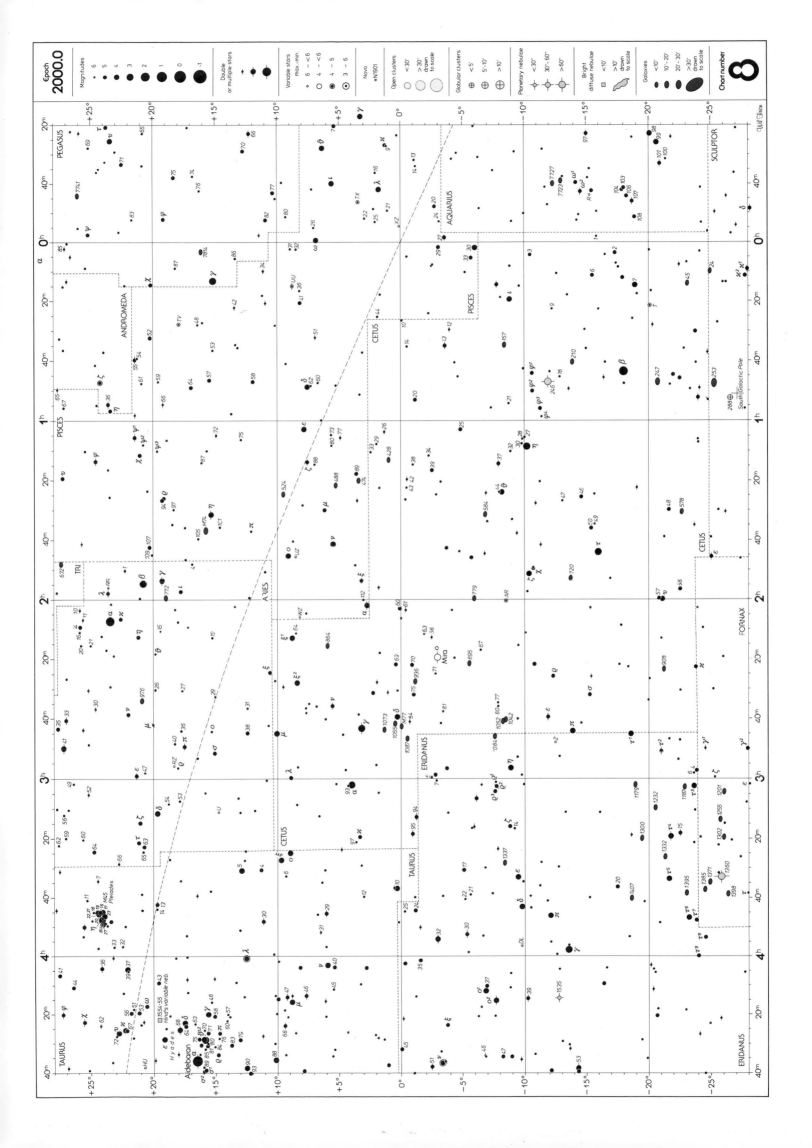

Chart 9. RA 4h to 8h. Dec +20° to −20°

Variable stars

		RA (h m)	Dec (°)	Range	Type	Period (d)	Spectrum
R	CMa	07 19.5	−16 24	5.7– 6.3	EA	1.14	F
R	Lep	04 59.6	−14 48	5.5–11.7	M	432.1	N
RX	Lep	05 11.4	−11 51	5.0– 7.0	Irr	–	M
U	Mon	07 30.8	−09 47	6.1– 8.1	RV Tau	92.3	F-K
V	Mon	06 22.7	−02 12	6.0–13.7	M	333.8	M
α	Ori	05 55.2	+07 24	0.1– 0.9	SR	2110	M
VV	Ori	05 33.5	−01 09	5.1– 5.5	EB	1.49	B+A
W	Ori	05 05.4	+01 11	5.9– 7.7	SR	212	N
BL	Ori	06 25.5	+14 43	8.5– 9.7	Irr	–	N
CK	Ori	05 30.3	+04 12	5.9– 7.1	SR	120	K
λ	Tau	04 00.7	+12 29	3.3– 3.8	EA	3.95	B+A
CE	Tau	05 32.2	+18 36	6.1– 6.5	SR	165	M

Double stars

		RA	Dec	PA	Sep	Mag	
α	CMa	06 45.1	−16 43	005	4.5	−1.5, 8.5	Sirius
μ	CMa	06 56.1	−14 03	{ AB 340	3.0	5.3, 8.6	
				AC 288	88.4	10.5	
				AD 061	101.3	10.7	
ν'	CMa	06 36.4	−18 40	262	17.5	5.8, 8.5	
α	CMi	07 39.3	+05 14	021	5.2	0.4,12.9	
η	CMi	07 28.0	+06 57	025	4.0	5.3,11.1	
o²	Eri	04 15.2	−07 39	104	83.4	4.4, 9.5	
53	Eri	04 38.2	−14 18	029	0.7	4.0, 7.0	
54	Eri	04 40.4	−19 40	161	0.3	4.9, 5.2	
55	Eri	04 43.6	−08 48	317	9.2	6.7, 6.8	
66	Eri	05 06.8	−04 39	009	52.8	5.2, 8.4	
λ	Gem	07 18.1	+16 32	033	9.6	3.6,10.7	
38	Gem	06 54.6	+13 11	147	7.0	4.7, 7.7	Binary, 3190 y
α	Lep	05 32.7	−17 49	{ AB 156	35.8	2.6,11.1	
				AC 186	91.4	11.8	
ι	Lep	05 12.3	−11 52	337	12.7	4.5,10.8	
κ	Lep	05 13.2	−12 56	358	2.6	4.5, 7.4	
β	Mon	06 28.8	−07 02	{ AB 132	7.3	4.7, 5.2	
				AC 124	100	6.1	
				AD 056	25.9	12.2	
γ	Mon	06 14.9	−06 16	027	51.4	4.0,13.0	
ε	Mon	06 23.8	+04 36	027	13.4	4.5, 6.5	
S(15)	Mon	06 41.0	+09 54	{ AB 213	2.8	4.7, 7.5	
				AC 013	16.6	9.8	
				AD 308	41.3	9.6	
				AE 139	73.9	9.9	
				AF 222	156.0	7.7	
				AK 056	105.6	8.1	
β	Ori	05 14.5	−08 12	202	9.5	0.1, 6.8	Rigel
ζ	Ori	05 40.8	−01 57	{ AB 162	2.4	1.9, 4.0	Binary, 1509 y
				AC 010	57.6	9.9	
η	Ori	05 24.5	−02 24	{ AC 080	1.5	3.8, 4.8	
				AC 051	115.1	9.4	
θ	Ori	05 35.3	−05 23	{ AB 031	8.8	6.7, 7.9	Trapezium
				AC 132	12.8	5.1	
				AD 096	21.5	6.7	
ι	Ori	05 35.4	−05 55	141	11.3	2.8, 6.9	
μ	Ori	06 02.4	+09 39	AB 023	0.4	4.4, 6.0	
π³	Ori	04 49.8	+06 58	138	94.6	3.2, 8.7	
ρ	Ori	05 13.3	+02 54	064	7.0	4.5, 8.3	
14	Ori	05 07.9	−08 30	349	0.7	5.8, 6.5	Binary, 199 y
75	Ori	06 17.1	+09 57	{ AB 258	62.7	5.4, 9.5	
α	Tau	04 35.9	+16 31	{ AB 110	30.4	0.9,13.4	Optical
				AC 034	121.7	11.1	
θ	Tau	04 28.7	+15 32	346	337.4	3.4, 3.8	
σ	Tau	04 39.3	+15 55	193	431.2	4.7, 5.1	
47	Tau	04 13.9	+09 16	351	1.1	4.9, 7.4	
66	Tau	04 23.9	+09 28	265	0.1	5.8, 5.9	Binary, 51.6 y
126	Tau	05 41.3	+16 32	238	0.3	5.3, 5.9	

Open clusters

NGC		RA	Dec	Diam	Mag	N*	
—	Tau	04 27	+16	330	1	200+	Hyades
1647	Tau	04 46.0	+19 04	45	6.4	200	
1807	Tau	05 10.7	+16 32	17	7.0	20	Asterism?
1817	Tau	05 12.1	+16 42	16	7.7	60	
1981	Ori	05 35.2	−04 26	25	4.6	20	
2112	Ori	05 53.9	+00 24	11	9.1	50	
2169	Gem	06 08.4	+13 57	7	5.9	30	
2186	Ori	06 12.2	+05 27	4	8.7	30	
2215	Mon	06 21.0	−07 17	11	8.4	40	
2244	Mon	06 32.4	+04 52	24	4.8	100	In Rosette Nebula
2251	Mon	06 34.7	+08 22	10	7.3	30	
2286	Mon	06 47.6	−03 10	15	7.5	50	
2301	Mon	06 51.8	+00 28	12	6.0	80	
2323	Mon	07 03.2	−08 20	16	5.9	80	M50
2335	Mon	07 06.6	−10 05	12	7.2	35	
2343	Mon	07 08.3	−10 39	7	6.7	20	
2345	CMa	07 08.3	−13 10	12	7.7	20	
2353	Mon	07 14.6	−10 18	20	7.1	30	
2355	Gem	07 16.9	+13 47	9	9.7	40	
2360	CMa	07 17.8	−15 37	13	7.2	80	
2395	Gem	07 27.1	+13 35	12	8.0	30	Asterism?
2422	Pup	07 36.6	−14 30	30	4.4	30	M47
2437	Pup	07 41.8	−14 49	27	6.1	100	M46 Contains p.n. NGC 2438
2479	Pup	07 55.1	−17 43	7	9.6	45	
Mel 71	Pup	07 37.5	−12 04	9	7.1	80	

Planetary nebulae

NGC		RA	Dec	Diam	Mag	Mag*	
1535	Eri	04 14.2	−12 44	18×44	9.6	12.2	
2438	Pup	07 41.8	−14 44	66	10.1	17.7	In cluster NGC 2437
IC 418	Lep	05 27.5	−12 42	12	10.7	10.7	

Nebulae

NGC		RA	Dec	Diam	Mag*	
1554–5	Tau	04 21.8	+19 32	var	9v	Hind's Variable Nebula (T Tau)
1976	Ori	05 35.4	−05 27	66×60	5	M42
1982	Ori	05 35.6	−05 16	20×15	7	M43 Extension of M42
2068	Ori	05 46.7	+00 03	8× 6	10	M78 Nebula is mag 8
2149	Mon	06 03.5	−09 44	3× 2	9	
2237–9	Mon	06 32.3	+05 03	80×60	—	Rosette Nebula. Contains cluster NGC 2244
2261	Mon	06 39.2	+08 44	2× 1	10v	R Mon: Hubble's Variable Nebula
2264	Mon	06 40.9	+09 54	60×30	4v	S Mon: Cone Nebula
I.C.434	Ori	05 41.0	−02 24	60×10	2(ζ)	Behind Horse's Head Nebula, Barnard 33

Galaxy

NGC		RA	Dec	Mag	Diam	Type
1637	Eri	04 41.5	−02 51	10.9	3.3×2.9	Sc

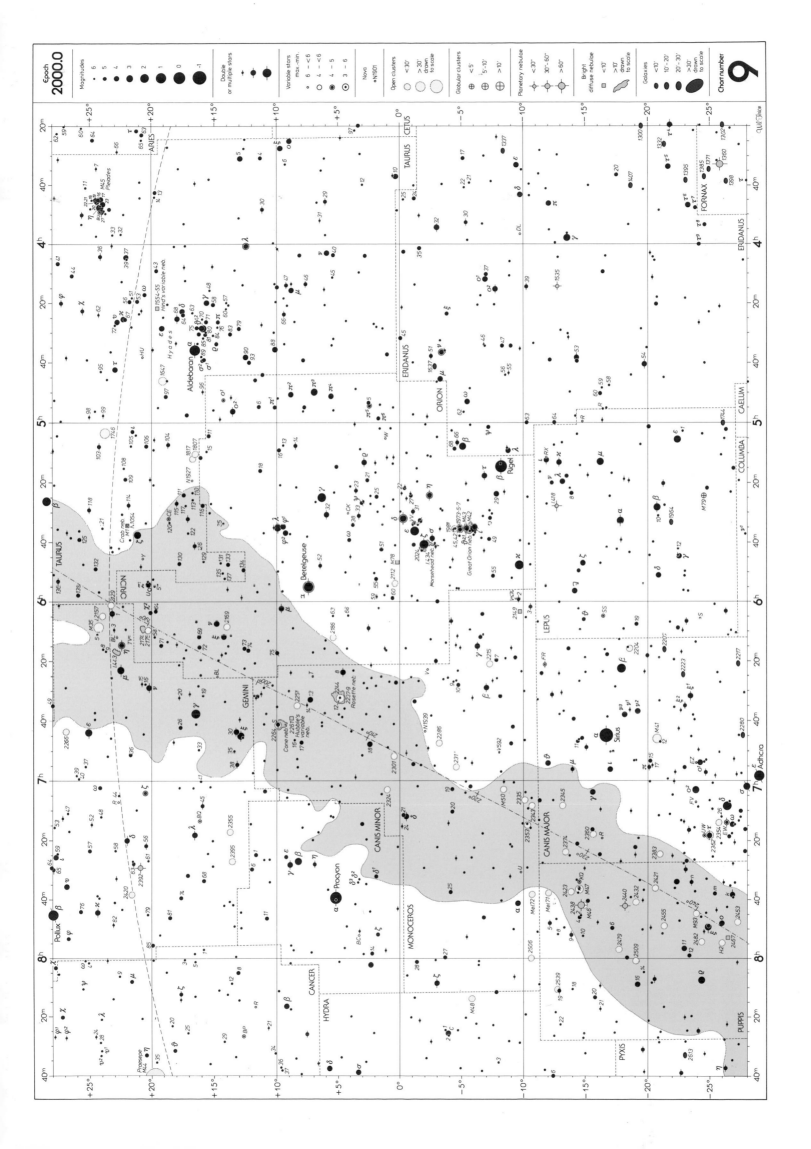

Chart 10. RA 8h to 12h. Dec +20° to −20°

Variable stars

	RA h m	Dec °	Range	Type	Period d	Spectrum
R Cnc	08 16.6	+11 44	6.1–11.8	M	361.6	M
X Cnc	08 55.4	+17 14	5.6– 7.5	SR	195	N
IN Hya	09 20.6	+00 11	6.3– 6.7	SR	45	M
R Leo	09 47.6	+11 25	4.4–11.3	M	312.4	M
VY Leo	10 56.0	+06 11	5.7– 6.0	Irr	—	M

Double stars

	RA h m	Dec °	PA	Sep	Mag
δ Cnc	08 44.7	+18 09	090	38.4	3.9,11.9 Optical
ζ Cnc	08 12.2	+17 39	{AB+C 088 / AB 182 / AB+D 108}	5.7 / 0.6 / 287.9	5.0, 6.2 Binary, 1150 y / 5.3, 6.0 Binary, 59.7 y / 9.7
γ Crt	11 24.9	−17 41	096	5.2	4.1, 9.6
ι Crt	11 38.7	−13 12	226	1.4	5.5,10.9
α Hya	09 27.6	−08 40	153	283.1	2.0, 9.5
ε Hya	08 46.8	+06 25	{AB 295 / AB+C 281}	0.2 / 2.8	3.8, 4.7 Binary, 890 y / 6.8
θ Hya	09 14.4	+02 19	197	29.4	3.9, 9.9
27 Hya	09 20.5	−09 33	211	229.4	5.0, 6.9
29 Hya	09 27.2	−09 13	{AB 184 / AB+C 175}	0.2 / 10.8	7.2, 7.3 / 11.8
α Leo	10 08.4	+11 58	307	176.9	1.4, 7.7 Regulus
γ Leo	10 20.0	+19 51	{AB 124 / AC 291 / AD 302}	4.3 / 259.9 / 333.0	2.2, 3.5 Binary, 619 y / 9.2 / 9.6
ι Leo	11 23.9	+10 32	131	1.5	4.0, 6.7 Binary, 192 y
ω Leo	09 28.5	+09 03	053	0.5	5.9, 6.5 Binary, 118 y
χ Leo	11 05.0	+07 20	{AB 262 / AC 305}	3.3 / 276.4	4.6,10.9 / 8.9
TX Leo	10 35.0	+08 39	157	2.4	5.8, 8.5
3 Leo	09 28.5	+08 11	080	25.2	5.7,10.4
6 Leo	09 32.0	+09 43	075	37.4	5.2, 8.2
31 Leo	10 07.9	+10 00	044	7.9	4.4,13.4
34 Leo	10 11.6	+13 21	286	0.2	6.8, 7.6
90 Leo	11 34.7	+16 48	{AB 209 / AC 234}	3.3 / 63.1	6.0, 7.3 / 8.7
γ Sex	09 52.5	−08 06	{AB 067 / AC 325}	0.6 / 35.8	5.6, 6.1 Binary, 75.6 y / 12.0

Open clusters

NGC	RA	Dec	Diam	Mag	N*	
2506 Mon	08 00.2	−10 47	7	7.6	150	
2509 Pup	08 00.7	−19 04	8	9.3	70	
2539 Pup	08 10.7	−12 50	22	6.5	50	
2548 Hya	08 13.8	−05 48	54	5.8	80	M48
2632 Cnc	08 40.1	+19 59	95	3.1	50	M44 Praesepe
2682 Cnc	08 50.4	+11 49	30	6.9	200	M67

Planetary nebula

NGC	RA	Dec	Diam	Mag	Mag*	
3242 Hya	10 24.8	−18 38	16×1250	8.6	12.0	Ghost of Jupiter

Galaxies

NGC	RA	Dec	Mag	Diam	Type	
2775 Cnc	09 10.3	+07 02	10.3	4.5×3.5	Sa	
2967 Sex	09 42.1	+00 20	11.6	3.0×2.9	Sc	
3115 Sex	10 05.2	−07 43	9.1	8.3×3.2	E6	
3166 Sex	10 13.8	+03 26	10.6	5.2×2.7	SBa	
3169 Sex	10 14.2	+03 28	10.4	4.8×3.2	Sb	
3351 Leo	10 44.0	+11 42	9.7	7.4×5.1	SBb	M95
3368 Leo	10 46.8	+11 49	9.2	7.1×5.1	Sb	M96
3377 Leo	10 47.7	+11 59	10.2	4.4×2.7	E5	
3379 Leo	10 47.8	+12 35	9.3	4.5×4.0	E1	M105
3384 Leo	10 48.3	+12 38	10.0	5.9×2.6	E7	
3412 Leo	10 50.9	+13 25	10.6	3.6×2.0	E5	
3489 Leo	11 00.3	+13 54	10.3	3.7×2.1	E6	
3521 Leo	11 05.8	−00 02	8.9	9.5×5.0	Sb	
3571 Crt	11 11.5	−18 17	12.8	3.3×1.3	Sa	
3593 Leo	11 14.6	+12 49	11.0	5.8×2.5	Sb	
3596 Leo	11 15.1	+14 47	11.6	4.2×4.1	Sc	
3607 Leo	11 16.9	+18 03	10.0	3.7×3.2	E1	
3623 Leo	11 18.9	+13 05	9.3	10.0×3.3	Sb	M65
3626 Leo	11 20.1	+18 21	10.9	3.1×2.6	Sb	
3627 Leo	11 20.2	+12 59	9.0	8.7×4.4	Sb	M66
3628 Leo	11 20.3	+13 36	9.5	14.8×3.6	Sb	Arp 317
3630 Leo	11 20.3	+02 58	12.8	2.3×0.9	E7	
3640 Leo	11 21.1	+03 14	10.3	4.1×3.4	E1	
3672 Crt	11 25.0	−09 48	11.5	4.1×2.1	Sb	
3686 Leo	11 27.7	+17 13	11.4	3.3×2.6	Sc	
3810 Leo	11 41.0	+11 28	10.8	4.3×3.1	Sc	
3887 Crt	11 47.1	−16 51	11.0	3.3×2.7	Sc	
3981 Crt	11 56.1	−19 54	12.4	3.9×1.5	Sb	

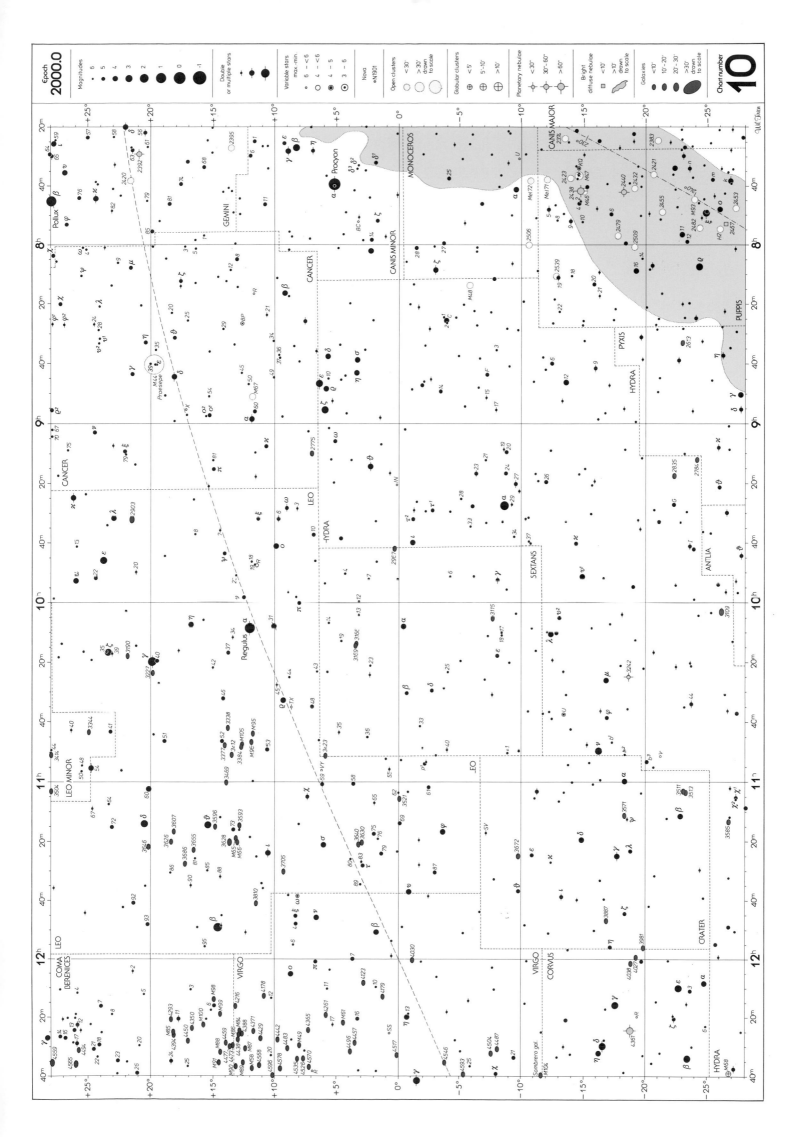

Chart 11. RA 12h to 16h. Dec +20° to −20°

Variable stars

		RA h m	Dec °	Range	Type	Period d	Spectrum
R	Crv	12 19.6	−19 15	6.7–14.4	M	317.0	M
R	Ser	15 50.7	+15 08	5.1–14.4	M	356.4	M
δ	Lib	15 01.1	−08 31	4.9– 5.9	EA	2.33	B
R	Vir	12 38.5	+06 59	6.0–12.1	M	145.6	M
S	Vir	13 33.0	−07 12	6.3–13.2	M	377.4	M
SS	Vir	12 25.3	+00 48	6.0– 9.6	M	354.7	N

Double stars

		RA	Dec	PA	Sep	Mag	
ζ	Boö	14 41.1	+13 44	{AB 303	1.0	4.5, 4.6	Binary, 123.3 y
				{AC 259	99.3	10.9	
π	Boö	14 40.7	+16 25	108	5.6	4.9, 5.8	
24	Com	12 35.1	+18 23	271	20.3	5.2, 6.7	
δ	Crv	12 29.9	−16 31	214	24.2	3.0, 9.2	
α	Lib	14 50.9	−16 02	α²–α¹ 314	231.0	2.8, 5.2	
ι	Lib	15 12.2	−19 47	111	57.8	5.1, 9.4	
κ	Lib	15 41.9	−19 41	279	172.0	4.7, 9.7	
μ	Lib	14 49.3	−14 09	{AB 355	1.8	5.8, 6.7	
				{AC 289	15.0	14.5	
				{AD 174	25.0	13.9	
18	Lib	14 58.9	−11 09	{AB 039	19.7	5.8,10.0	
				{AC 41	162.3	11.3	
47	Lib	15 55.0	−19 23	129	0.5	6.1, 8.1	
β	Ser	15 46.2	+15 25	265	30.6	3.7, 9.9	
δ	Ser	15 34.8	+10 32	177	4.4	4.1, 5.2	Binary, 3168 y
5	Ser	15 19.3	+01 46	{AB 036	11.2	5.1,10.1	
				{AC 040	127.2	9.1	
6	Ser	15 21.0	+00 43	020	3.1	5.4,10.0	
γ	Vir	12 41.7	−01 27	287	3.0	3.5, 3.5	Binary, 171.4 y
θ	Vir	13 09.9	−05 32	343	7.1	4.4, 9.4	
τ	Vir	14 01.6	+01 33	290	80.0	4.3, 9.6	
17	Vir	12 22.5	+05 18	337	20.0	6.6, 9.4	
31	Vir	12 42.0	+06 48	037	4.0	5.6,11.4	
73	Vir	13 32.0	−18 44	183	0.1	6.7, 6.9	
84	Vir	13 43.1	+03 32	229	2.9	5.5, 7.9	

Open cluster

NGC		RA	Dec	Diam	Mag	N*	
Mel 111	Com	12 25	+26	275	4	80	Coma Berenices

Globular clusters

NGC		RA	Dec	Diam	Mag	
5024	Com	13 12.9	+18 10	1	7.7	M53
5904	Ser	15 18.6	+02 05	17	5.8	M5

Planetary nebula

NGC		RA	Dec	Diam	Mag	Mag*
4631	Crv	12 24.5	−18 48	45×110	10.3	13.2

Galaxies*

NGC		RA	Dec	Mag	Diam	Type	
4192	Com	12 13.8	+14 54	10.1	9.5×3.2	Sb	M98
4216	Vir	12 15.9	+13 09	10.0	8.3×2.2	Sb	
4254	Com	12 18.8	+14 25	9.8	5.4×4.8	Sc	M99
4261	Vir	12 19.4	+05 49	10.3	3.9×3.2	E2	
4303	Vir	12 21.9	+04 28	9.7	6.0×5.5	Sc	M61
4321	Com	12 22.9	+15 49	9.4	6.9×6.2	Sc	M100
4374	Vir	12 25.1	+12 53	9.3	5.0×4.4	E1	M84
4429	Vir	12 27.4	+11 07	10.2	5.5×2.6	SO	
4438	Vir	12 27.8	+13 01	10.1	9.3×3.9	Sap	
4442	Vir	12 28.1	+09 48	10.5	4.6×1.9	E5p	
4450	Com	12 28.5	+17 05	10.1	4.8×3.5	Sb	
4406	Vir	12 26.2	+12 57	9.2	7.4×5.5	Sb	M86
4459	Com	12 29.0	+13 59	10.4	3.8×2.8	E2	
4472	Vir	12 29.8	+08 00	8.4	8.9×7.4	E4	M49
4473	Com	12 29.8	+13 26	10.2	4.5×2.6	E4	
4477	Com	12 30.0	+13 38	10.4	4.0×3.5	SBa	
4486	Vir	12 30.8	+12 24	8.6	7.2×6.8	E1	M87 Virgo A
4501	Com	12 32.0	+14 25	9.5	6.9×3.9	Sb	M88
4535	Vir	12 34.3	+08 12	9.8	6.8×5.0	SBc	
4546	Vir	12 35.5	−03 48	10.3	3.5×1.7	E6	
4548	Com	12 35.4	+14 30	10.2	5.4×4.4	SBb	M91
4552	Vir	12 35.7	+12 33	9.8	4.2×4.2	E0	M89
4569	Vir	12 36.8	+13 10	9.5	9.5×4.7	Sb	M90
4579	Vir	12 37.7	+11 49	9.8	5.4×4.4	Sb	M58
4594	Vir	12 40.0	−11 37	8.3	8.9×4.1	Sb	M104 Sombrero Hat
4596	Vir	12 39.9	+10 11	10.5	3.9×2.8	SBa	
4621	Vir	12 42.0	+11 39	9.8	5.1×3.4	E3	M59
4636	Vir	12 42.8	+02 41	9.6	6.2×5.6	E1	
4649	Vir	12 43.7	+11 33	8.8	7.2×6.2	E1	M60
4651	Com	12 43.7	+16 24	10.7	3.8×2.7	Scp	
4654	Vir	12 44.0	+13 08	10.5	4.7×3.0	Sc	
4689	Com	12 47.8	+13 46	10.9	4.0×3.5	Sc	
4697	Vir	12 48.6	−05 48	9.3	6.0×3.8	E4	
4699	Vir	12 49.0	−08 40	9.6	3.5×2.7	Sa	
4753	Vir	12 52.4	−01 12	9.9	5.4×2.9	Pec	
4762	Vir	12 52.9	+11 14	10.2	8.7×1.6	SB0	
4856	Vir	12 59.3	−15 02	10.4	4.6×1.6	SBa	
5247	Vir	13 38.1	−17 53	10.5	5.4×4.7	Sb	
5248	Boö	13 37.5	+08 53	10.2	6.5×4.9	Sc	
5363	Vir	13 56.1	+05 15	10.2	4.2×2.7	Ep	
5364	Vir	13 56.2	+05 01	10.4	7.1×5.0	SB+p	

* There are so many galaxies in this region that a selection has had to be made, with a general limiting magnitude of 10.5.

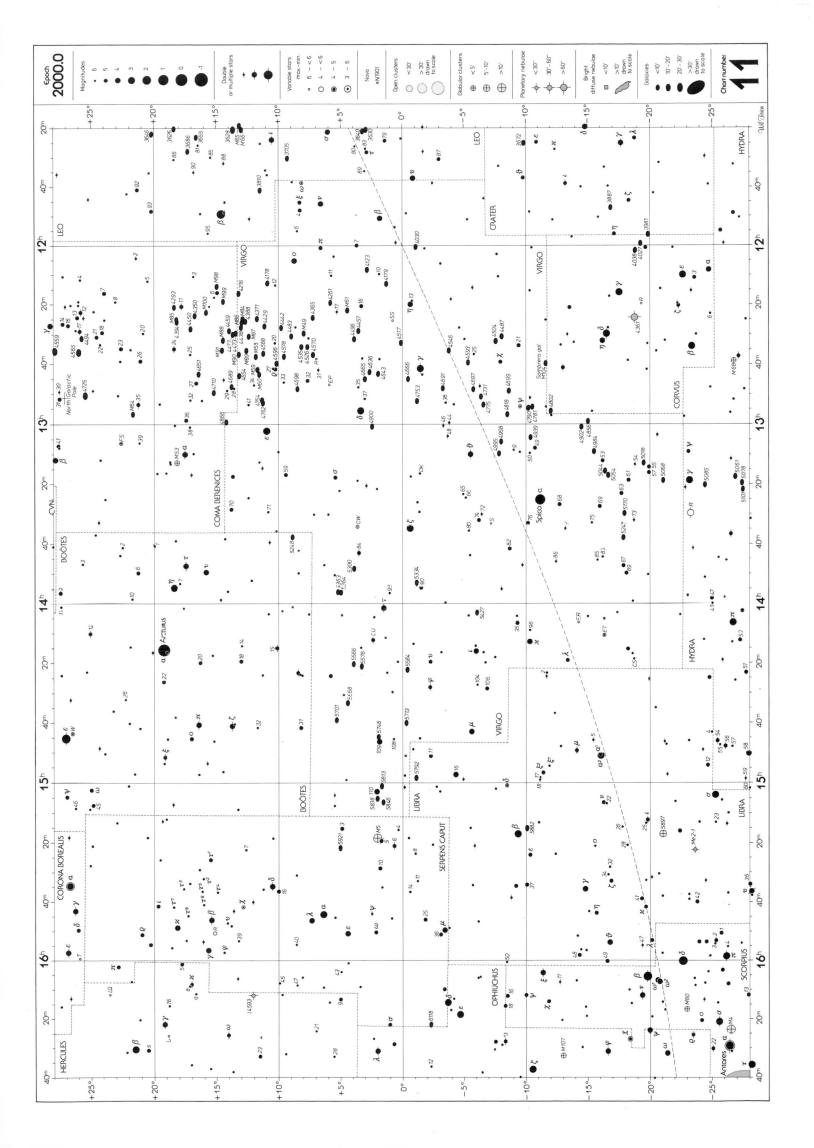

Chart 12. RA 16h to 20h. Dec +20° to −20°

Variable stars

		RA h m	Dec °	Range	Type	Period d	Spectrum
η	Aql	19 52.5	+01 00	3.5– 4.4	Cep	7.18	F–G
R	Aql	19 06.4	+08 14	5.5–12.0	M	284.2	M
U	Aql	19 29.4	−07 03	6.1– 6.9	Cep	7.02	F–G
V	Aql	19 04.4	−05 41	6.6– 8.4	SR	353	N
α	Her	17 14.6	+14 23	3.0– 4.0	SR	~100?	M
S	Her	16 51.9	+14 56	6.4–13.8	M	307.4	M
U	Her	16 25.8	+18 54	6.5–13.4	M	406.0	M
χ	Oph	16 27.0	−18 27	4.2– 5.0	Irr	–	B
U	Oph	17 16.5	+01 13	5.9– 6.6	EA	1.68	B+B
Y	Oph	17 52.6	−06 09	5.9– 6.4	Cep	17.12	F–G
R	Sct	18 47.5	−05 42	4.4– 8.2	RV Tau	140	G–K
S	Sge	19 56.0	+16 38	5.3– 6.0	Cep	8.38	F–G
U	Sge	19 18.8	+19 37	6.6– 9.2	EA	3.38	B+K

Double stars

		RA h m	Dec °	Mag	Sep	PA
α	Aql	19 50.8	+08 52	0.8, 9.5 Altair	165.2	301
γ	Aql	19 46.3	+10 37	2.7,10.7	132.6	238
δ	Aql	19 25.5	+03 07	3.4,10.9	108.9	271
ε	Aql	18 59.6	+15 04	4.0, 9.9	131.1	187
ι	Aql	19 36.7	−01 17	4.4,13.1	47.0	161
μ	Aql	19 34.1	+07 23	4.5,13.0	31.2	280
ν	Aql	19 26.5	+00 20	4.7, 8.9	201.0	288
π	Aql	19 48.7	+11 49	6.1, 6.9	1.4	110
χ	Aql	19 42.6	+11 50	5.6, 6.8	0.5	077
U	Aql	19 29.4	−07 03	var,11.7	1.5	228
23	Aql	19 18.5	+01 05	5.3, 9.3	3.1	005
31	Aql	19 25.0	+11 57	5.2, 8.7	105.6	343
57	Aql	19 54.6	−08 14	5.8, 6.5	35.7	170
α	Her	17 14.6	+14 23	var, 5.4 Binary, 3600 y	4.7	107
κ	Her	16 08.1	+17 03	5.3, 6.5	28.4	012
ω	Her	16 25.4	+14 02	4.6,11.6 / 11.1	1.0 / 28.4	{AB 223 / AC 096}
37	Her	16 40.6	+04 13	5.8, 7.0	69.8	230
54	Her	16 55.4	+18 26	5.4,12.7	2.5	183
η	Oph	17 10.4	−15 43	3.0, 3.5 Binary, 84.3 y	0.5	247
λ	Oph	16 30.9	+01 59	4.2, 5.2 Binary, 129.9 y / 11.1 / 9.9	1.5 / 119.2 / 313.8	{AB 022 / AB+C 170 / AD 246}
τ	Oph	18 03.1	−08 11	5.2, 5.9 Binary, 280 y / 9.3	1.8 / 100.3	{AB 280 / AC 127}
ν	Oph	16 27.8	−08 22	4.6, 7.8	1.0	095
φ	Oph	16 31.1	−16 37	4.3,12.8	34.4	037
X	Oph	18 38.3	+08 50	5.9v, 8.6 Binary, 485 y	0.4	150
19	Oph	16 47.2	+02 04	6.1, 9.4	23.4	089
41	Oph	17 16.6	−00 27	4.8, 7.8	1.0	346
53	Oph	17 34.6	+09 35	5.8, 8.5	41.2	191
70	Oph	18 05.5	+02 30	4.2, 6.0 Binary, 88.1 y 4 faint comps	1.5	{AB 224}
73	Oph	18 09.6	+04 00	6.1, 7.0 Binary, 270 y	0.4	300
β	Sco	16 05.4	−19 48	2.6, 4.9 A is a close double	13.6	{AC 021}
ν	Sco	16 12.0	−19 28	4.3, 6.8 Binary, 45.7 y / 6.4	0.9 / 41.1	{AB 003 / AC 337}
ξ	Sco	16 04.4	−11 22	4.8, 5.1 / 7.3	0.8 / 7.6	{AB 040 / AC 051}
11	Sco	16 07.6	−12 45	5.6, 9.9	3.3	257
θ	Ser	18 56.2	+04 12	4.5, 4.5	22.3	104
ν	Ser	17 20.8	−12 51	4.3, 8.3	46.3	028
59	Ser (d)	18 27.2	+00 12	5.3, 7.6	3.8	318
α	Sge	19 40.1	+18 01	4.4,13.2 / 14.9	31.5	{AB 179 / AC 249}
ζ	Sge	19 49.0	+19 09	5.5, 8.7 / 5.5, 6.2 Binary, 22.8 y	35.8 / 0.3	{AB+C 311 / AB 163}

Open clusters

NGC		RA	Dec	Diam	Mag	N*	
M24	Sgr	18 16.9	−18 29	90	4.5	–	Not a true cluster. Star-cloud in the Milky Way
6494	Sgr	17 56.8	−19 01	27	5.5	150	M23
6613	Sgr	18 19.9	−17 08	9	6.9	20	M18
6633	Oph	18 27.7	+06 34	27	4.6	30	
6645	Sgr	18 32.6	−16 54	10	8.5	40	
6664	Sct	18 36.7	−08 13	16	7.8	50	EV Scuti cluster
6694	Sct	18 45.2	−09 24	15	8.0	30	
6704	Sct	18 50.9	−05 12	6	9.2	30	
6705	Sct	18 51.1	−06 16	14	5.8	500	M11 Wild Duck
6709	Aql	18 51.5	+10 21	13	6.7	40	
6716	Sgr	18 54.6	−19 53	7	6.9	20	
6755	Aql	19 07.8	+04 14	15	7.5	100	
IC 4665	Oph	17 46.3	+05 43	41	4.2	30	
IC 4725	Sgr	18 31.6	−19 15	32	4.6	31	M25: ν Sgr.
H.20	Sge	19 53.1	+18 20	7	7.7	15	

Globular clusters

NGC		RA	Dec	Diam	Mag	
6171	Oph	16 32.5	−13 03	10.0	8.1	M107
6218	Oph	16 47.2	−01 57	14.5	6.6	M12
6254	Oph	16 57.1	−04 06	15.1	6.6	M10
6333	Oph	17 19.2	−18 31	9.3	7.9	M9
6356	Oph	17 23.6	−17 49	7.2	8.4	
6402	Oph	17 37.6	−03 15	11.7	7.6	M14
6712	Sct	18 53.1	−08 42	7.2	8.2	
6838	Sge	19 53.8	+18 47	7.2	8.3	M71

Planetary nebulae

NGC		RA	Dec	Diam	Mag	Mag*
6572	Oph	18 12.1	+06 51	8	9.0	13.6
6741	Aql	19 02.6	−00 27	6	10.8	14.7
6751	Aql	19 05.9	−06 00	20	12.5	13.9
6790	Aql	19 23.2	+01 31	7	10.2	13.5
6818	Sgr	19 44.0	−14 09	17	9.9	13.0
IC 4593	Her	16 12.2	+12 04	12×120	10.9	11.3

Nebulae

NGC		RA	Dec	Diam	Mag*	
6611	Ser	18 18.8	−13 47	35×28	—	M16 Eagle Nebula, with cluster
6618	Sgr	18 20.8	−16 11	46×37	—	M17 Omega Nebula

Galaxies

NGC		RA	Dec	Diam	Mag	Type	
6118	Ser	16 21.8	−02 17	4.7×2.3	12.3	Sb	
6384	Oph	17 32.4	+07 04	6.0×4.3	10.6	Sb	
6822	Sgr	19 44.9	−14 48	10.2×7.5	9.3	Irr	Barnard's Galaxy

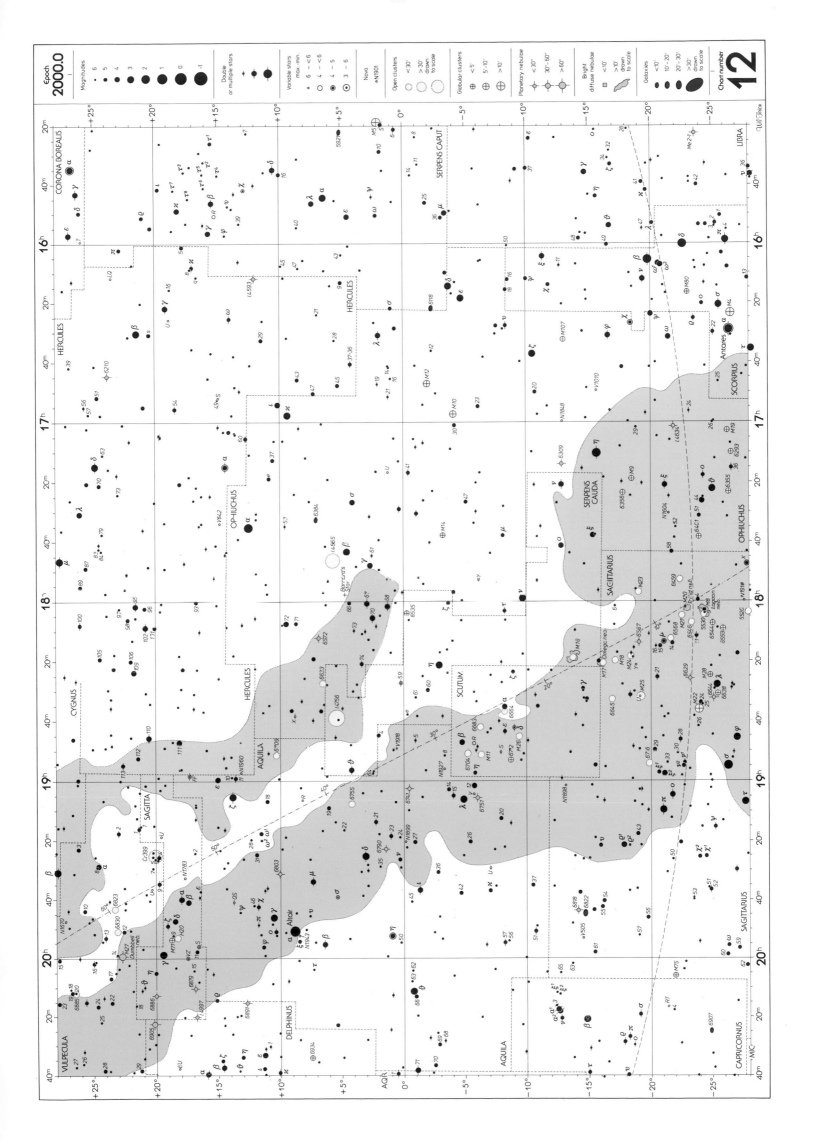

Chart 13. RA 20h to 0h. Dec +20° to −20°

Variable stars

	RA h m	Dec °	Range	Type	Period d	Spectrum
R Aqr	23 43.8	−15 17	5.8–12.4	Symbiotic	387.0	M+Pec
DV Aqr	20 58.7	−14 29	5.9– 6.2	EB	1.58	F
U Del	20 45.5	+18 05	7.6– 8.9	SR	110	M
EU Del	20 37.9	+18 16	5.8– 6.9	SR	59	M
TX Psc	23 46.4	+03 29	6.9– 7.7	Irr	–	N
XZ Psc	23 54.8	+00 07	5.5– 6.0	Irr	–	M
VZ Sge	20 00.1	+17 31	5.3– 5.6	Irr	–	M
AG Peg	21 51.0	+12 38	6.0– 9.4	Z And	830	WN+M

Double stars

	RA	Dec	PA	Sep	Mag	
68 Aql	20 28.4	−03 21	178	9.8	6.1,13.8	
β Aqr	21 31.6	−05 34	{ AB 321	35.4	2.9,10.8	
			AC 186	57.2	11.4	
ξ Aqr	22 28.8	−00 01	200	2.0	4.3, 4.5	Binary, 856 y
ψ¹ Aqr	23 15.9	−09 05	274	80.4	4.5,13.5	
ψ³ Aqr	23 19.0	−09 37	174	1.5	5.0,11.0	
51 Aqr	22 24.1	−04 50	{ AB 324	0.5	6.5, 6.5	
			AB+D 191	116.0	10.1	
			AC 342	54.4	10.2	
			AE 133	132.4	8.6	
107 Aqr	23 46.0	−18 41	136	6.6	5.7, 6.7	
α Cap	20 18.1	−12 33	α¹–α² 291	377.7	3.6, 4.2	
α¹ Cap	20 17.6	−12 30	{ AB 182	44.3	4.2,13.7	
			AC 221	45.4	9.2	
α² Cap	20 18.1	−12 33	{ AB 172	6.6	3.6,11.0	
			AD 156	154.6	9.3	
			BC 240	1.2	11.3	
ε Cap	21 37.1	−19 28	047	68.1	4.7, 9.5	
π Cap	20 27.3	−18 13	148	3.2	5.3, 8.9	
ρ Cap	20 28.9	−17 49	158	0.5	5.0,10.0	
σ Cap	20 19.6	−19 07	179	55.9	5.5, 9.0	
τ Cap	20 39.3	−14 57	118	0.3	5.8, 6.3	Binary, 200 y
α Del	20 39.6	+15 55	224	29.5	3.8,13.3	
β Del	20 37.5	+14 36	167	0.3	4.0, 4.9	Binary, 26.7 y
γ Del	20 46.7	+16 07	268	9.6	4.5, 5.5	
κ Del	20 39.1	+10 05	286	28.8	5.1,11.7	
1 Del	20 30.3	+10 54	{ AB 346	0.9	6.1, 8.1	
			AC 349	16.8	14.1	
13 Del	20 47.8	+06 00	194	1.6	5.6, 9.2	
β Eql	21 22.9	+06 49	257	34.4	5.2,13.7	
γ Eql	21 10.4	+10 08	{ AB 268	1.9	4.7,11.5	
			AC 005	47.7	12.5	
δ Eql	21 14.5	+10 00	029	0.3	5.2, 5.3	Binary, 5.7 y
ε Eql	20 59.1	+04 18	{ AB 285	1.0	6.0, 6.3	Binary, 101.4 y
			AB+C 070	10.7	7.1	
ε Peg	21 44.2	+09 52	AD 280	74.8	12.4	
ξ Peg	22 46.7	+12 10	{ AB 325	81.8	2.4,11.2	
			AC 320	142.5	8.4	
			{ AB 100	11.5	4.2,12.2	
			AC 015	145.0	11.0	
20 Peg	22 01.1	+13 07	324	54.7	5.7,11.1	
34 Peg	22 26.6	+04 24	{ AB 224	3.5	5.8,12.3	
			AC 272	103.3	12.8	
35 Peg	22 27.9	+04 42	{ AB 210	98.3	4.8, 9.8	
			AC 241	181.5	9.7	
37 Peg	22 30.0	+04 26	118	0.9	5.8, 7.1	Binary, 140 y
66 Peg	23 23.1	+12 19	056	0.1	5.9, 5.9	
2 Psc	22 59.5	+00 58	085	3.8	5.4,13.1	
15 Sge	20 04.1	+17 04	{ AB 276	190.7	5.9, 8.1	
			AC 320	203.7	6.8	

Open cluster

NGC	RA	Dec	Diam	Mag	N*
6994 Aqr	20 58.9	−12 38	2.8	8.9	4 M73 Asterism

Globular clusters

NGC	RA	Dec	Diam	Mag	
6934 Del	20 34.2	+07 24	5.9	8.9	
6981 Aqr	20 53.5	−12 32	5.9	9.3	M72
7078 Peg	21 30.0	+12 10	12.3	6.3	M15
7089 Aqr	21 33.5	−00 49	12.9	6.5	M2

Planetary nebulae

NGC	RA	Dec	Diam	Mag	Mag*
6879 Sge	20 10.5	+16 55	5	13.0	15
6891 Del	20 15.2	+12 42	12× 74	11.7	12.4
7009 Aqr	21 04.2	−11 22	25×100	8.3	11.5 Saturn Nebula
IC 4997 Sge	20 20.2	+16 45	2	11.6	13v?

Galaxies

NGC	RA	Dec	Mag	Diam	Type
7479 Peg	23 04.9	+12 19	11.0	4.1×3.2	SBb
7606 Aqr	23 19.1	−08 29	10.8	5.8×2.6	Sb
7723 Aqr	23 38.9	−12 58	11.1	3.6×2.6	Sb
7727 Aqr	23 39.9	−12 18	10.7	4.2×3.4	SBap

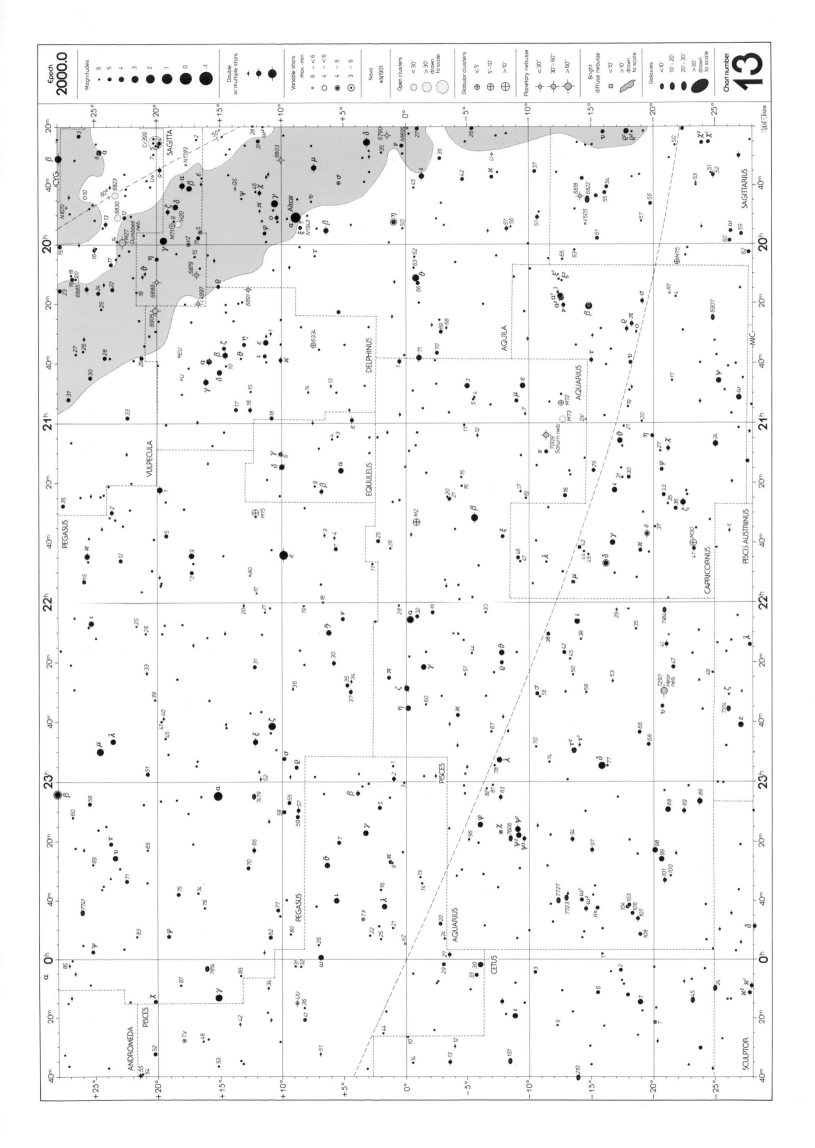

Chart 14. RA 0h to 4h. Dec −20° to −70°

Galaxies

NGC	RA	Dec	Mag	Diam	Type
24 Scl	00 09.9	−24 58	11.5	5.5 × 1.6	Sb
45 Cet	00 14.1	−23 11	10.4	8.1 × 5.8	S
55 Scl	00 14.9	−39 11	8.2	32.4 × 6.5	SB
134 Scl	00 30.4	−33 15	10.1	8.1 × 2.6	SBb
247 Cet	00 47.1	−20 46	8.9	20.0 × 7.4	S
253 Scl	00 47.6	−25 17	7.1	25.1 × 7.4	Scp
300 Scl	00 54.9	−37 41	8.7	20.0 × 14.8	Sd
578 Cet	01 30.5	−22 40	10.9	4.8 × 3.2	Sc
613 Scl	01 34.3	−29 25	10.0	5.8 × 4.6	SBb
685 Eri	01 47.8	−52 47	11.8	4.1 × 4.0	SBc
908 Cet	02 23.1	−21 14	10.2	5.5 × 2.8	Sc
986 For	02 33.6	−39 02	11.0	3.7 × 2.8	SBb
1097 For	02 46.3	−30 17	9.2	9.3 × 6.6	SBb
1187 Eri	03 02.6	−22 52	10.9	5.0 × 4.1	SBc
1201 For	03 04.1	−26 04	10.6	4.4 × 2.8	SO
1249 Hor	03 10.1	−53 21	11.7	5.2 × 2.7	SBc
1255 For	03 13.5	−25 44	11.1	4.1 × 2.8	Sa
1291 Eri	03 17.3	−41 08	8.5	10.5 × 9.1	SBa
1302 For	03 19.9	−26 04	11.5	4.4 × 4.2	SBa
1313 Ret	03 18.3	−66 30	9.4	8.5 × 6.6	SBd
1316 For	03 22.7	−37 12	8.8	7.1 × 5.5	SBOp
1326 For	03 23.9	−36 28	10.5	4.0 × 3.0	SBO
1332 Eri	03 26.3	−21 20	10.3	4.6 × 1.7	E7
1344 For	03 28.3	−31 04	10.3	3.9 × 2.3	E3
1350 For	03 31.1	−33 38	10.5	4.3 × 2.4	SBb
1365 For	03 33.6	−36 08	9.5	9.8 × 5.5	SBb
1371 For	03 35.0	−24 56	11.5	5.4 × 4.0	SBa
1380 For	03 36.5	−34 59	11.1	4.9 × 1.9	SO
1385 For	03 37.5	−24 30	11.2	3.0 × 2.0	Sc
1395 Eri	03 38.5	−23 02	11.3	3.2 × 2.5	E3
1398 For	03 38.9	−26 20	9.7	6.6 × 5.2	SBb
1399 For	03 38.5	−35 27	9.9	3.2 × 3.1	E1p
1404 For	03 38.9	−35 35	10.2	2.5 × 2.3	E1
1411 Hor	03 38.8	−44 05	11.9	2.8 × 2.3	SO
1425 For	03 42.2	−29 54	11.7	5.4 × 2.7	Sb
1433 Hor	03 42.0	−47 13	10.0	6.8 × 6.0	SBa
1448 Hor	03 44.5	−44 39	11.3	8.1 × 1.8	Sc
1493 Hor	03 57.5	−46 12	11.8	2.6 × 2.3	SBc

Variable stars

		RA h m	Dec °	Range	Type	Period d	Spectrum
R	Hor	02 53.9	−49 53	4.7–14.3	M	404.0	M
TW	Hor	03 12.6	−57 19	5.2– 5.9	SR	158	N
ζ	Phe	01 08.4	−55 15	3.9– 4.4	EA	1.67	B+B
R	Scl	01 27.0	−32 33	5.8– 7.7	SR	370	N
S	Scl	00 15.4	−32 03	5.5–13.6	M	365.3	M

Double stars

		RA	Dec	PA	Sep	Mag
θ	Eri	02 58.3	−40 18	088	8.2	3.4, 4.5
τ⁴	Eri	03 19.5	−21 45	{AB 288	5.7	3.7, 9.2
				{AC 112	39.2	10.5
χ	Eri	01 56.0	−51 37	202	5.0	3.7,10.7
p	Eri	01 39.8	−56 12	194	11.2	5.5, 5.8 Binary, 484 y
α	For	03 12.1	−28 59	298	4.0	4.0, 7.0 Binary, 314 y
γ¹	For	02 49.8	−24 34	{AB 145	12.0	6.1,12.5
				{AC 143	40.9	10.5
η²	For	02 50.2	−35 51	014	5.0	5.9,10.1
χ³	For	03 28.2	−35 51	248	6.3	6.5,10.5
ω	For	02 33.8	−28 14	244	10.8	5.0, 7.7
β	Phe	01 06.1	−46 43	346	1.4	4.0, 4.2
η	Phe	00 43.4	−57 28	217	19.8	4.4,11.4
ξ	Phe	00 41.8	−56 30	253	13.2	5.8,10.2
ε	Scl	01 45.6	−25 03	028	4.7	5.4, 8.6 Binary, 1192 y
ζ	Scl	00 02.3	−29 43	320	3.0	5.0,13.0
κ¹	Scl	00 09.3	−27 59	265	1.4	6.1, 6.2
λ¹	Scl	00 42.7	−38 28	003	0.7	6.7, 7.0
{β	Tuc	00 31.5	−62 58	169	27.1	4.4, 4.8
{β²	Tuc	00 31.6	−62 58	295	0.6	4.8, 6.0 Binary, 44.4 y
κ	Tuc	01 15.8	−68 53	336	5.4	5.1, 7.3

Globular clusters

NGC	RA	Dec	Mag	Diam
288 Scl	00 52.8	−26 35	8.1	13.8
1261 Hor	03 12.3	−55 13	8.4	6.9

Planetary nebula

NGC	RA	Dec	Mag	Diam	Mag*
1360 For	03 33.3	−25 51	—	390	11.3

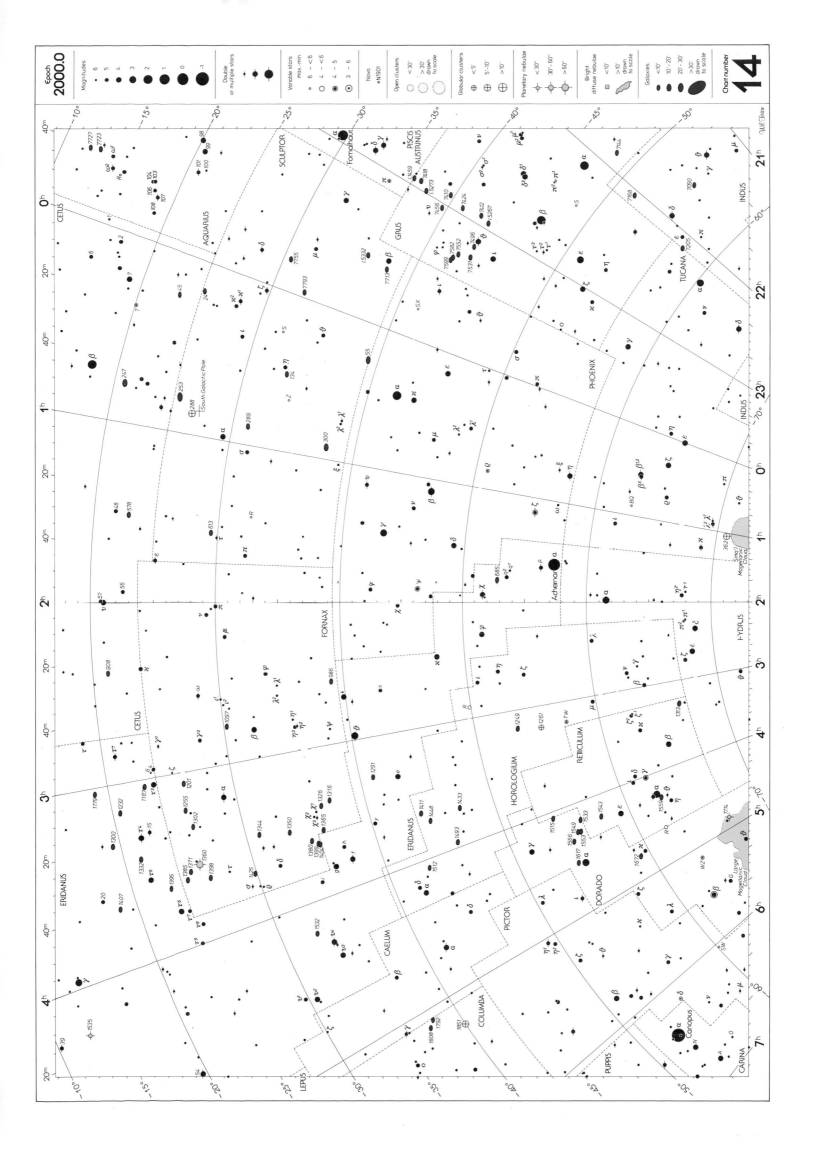

Chart 15. RA 4h to 8h. Dec −20° to −70°

Variable stars

	RA h m	Dec °	Type	Range	Period d	Spectrum
EW CMa	07 14.3	−26 21	Irr	4.4–4.8	—	B
UW CMa	07 18.7	−24 34	EB	4.8–5.3	4.39	O7
β Dor	05 33.6	−62 29	Cep	3.7–4.1	9.84	F–G
R Dor	04 36.8	−62 05	SR	4.8–6.6	338	M
L² Pup	07 13.5	−44 39	SR	2.6–6.2	140	M
V Pup	07 58.2	−49 15	EB	4.7–5.2	1.45	B+B

Double stars

	RA	Dec	PA	Sep	Mag	
α Cae	04 40.6	−41 52	121	6.6	4.5,12.5	
γ Cae	05 04.4	−35 29	308	2.9	4.6, 8.1	
ε CMa	06 58.6	−28 58	161	7.5	1.5, 7.4	
ξ¹ CMa	06 31.9	−23 25	144	24.6	4.3,13.9	
π CMa	06 55.6	−20 08	018	11.6	4.7, 9.7	
17 CMa	06 55.0	−20 24	AB 147	44.4	5.8, 9.3	
			AC 184	50.5	9.0	
			AD 186	129.9	9.5	
α Col	05 39.6	−34 04	359	13.5	2.6,12.3	
γ Col	05 57.5	−35 17	110	33.8	4.4,12.7	
π² Col	06 07.9	−42 09	150	0.1	6.2, 6.3	
α Dor	04 34.0	−55 03	AB 182	0.2	3.8, 4.3	
			AB+C 101	77.7	9.8	
υ⁴ Eri	04 17.9	−33 48	AB+C 013	49.2	3.6,11.8	A is a close double
β Lep	05 28.2	−20 46	AB 330	2.5	2.8, 7.3	
			AC 145	64.3	11.8	
			AD 075	206.4	10.3	
			AE 058	241.5	10.3	
η¹ Pic	05 02.8	−49 09	198	10.6	5.4,13.0	
θ Pic	05 24.8	−52 19	AB 152	0.2	6.9, 7.2	
			AB+C 287	38.2	6.8	
μ Pic	06 32.0	−58 45	231	2.4	5.8, 9.0	
σ Pup	07 29.2	−43 18	074	22.3	3.3, 9.4	
1 Pup	07 43.5	−28 24	033	26.2	4.6,13.5	

Open clusters

NGC	RA	Dec	Diam	Mag	N*	
2287 CMa	06 46.0	−20 44	38	4.5	80	M41
2362 CMa	07 17.8	−24 57	8	4.1	60	τ CMa cluster
2383 Pup	07 24.8	−20 56	6	8.4	40	
2421 Pup	07 36.3	−20 37	10	8.3	70	
2439 Pup	07 40.8	−31 39	10	6.9	80	R Pup Asterism
2447 Pup	07 44.6	−23 52	22	6.2	80	M93
2451 Pup	07 45.4	−37 58	45	2.8	40	
2455 Pup	07 49.0	−21 18	8	10.2	50	
2477 Pup	07 52.3	−38 33	27	5.8	160	
2489 Pup	07 56.2	−30 04	8	7.9	45	
2516 Car	07 58.3	−60 52	30	3.8	80	

Globular clusters

NGC	RA	Dec	Diam	Mag	
1851 Col	05 14.1	−40 03	11.0	7.3	X-ray source
1904 Lep	05 24.5	−24 33	8.7	8.0	M79

Nebulae

NGC	RA	Dec	Diam	Mag*	
2070 Dor	05 38.7	−69 06	40×25	—	30 Doradus in LMC
2467 Pup	07 52.5	−26 24	8× 7	9.2	Gum 9

Galaxies

NGC	RA	Dec	Mag	Diam	Type
1512 Hor	04 03.9	−43 21	10.6	4.0×3.2	SBa
1532 Eri	04 12.1	−32 52	11.1	5.6×1.8	Sb
1553 Dor	04 16.2	−55 47	9.5	4.1×2.8	SO
1559 Ret	04 17.6	−62 47	10.4	3.3×2.1	SBc
1549 Dor	04 15.7	−55 36	9.9	3.7×3.2	EO
1617 Dor	04 31.7	−54 36	10.4	4.7×2.4	SBa
1672 Dor	04 45.7	−59 15	11.0	4.8×3.9	SBb
1744 Lep	05 00.0	−26 01	11.2	6.8×4.1	SBc
1792 Col	05 05.2	−37 59	10.2	4.0×2.1	Sb
1808 Col	05 07.7	−37 32	9.9	7.2×4.1	SBa
1964 Lep	05 33.4	−21 57	10.8	6.2×2.5	Sb
2090 Col	05 47.0	−34 14	11.7	4.5×2.3	Sc
2207 CMa	06 16.4	−21 22	10.7	4.3×2.9	Sc
2217 CMa	06 21.7	−27 14	10.4	4.8×4.4	SBa
2223 CMa	06 24.6	−22 50	11.4	3.3×3.0	SBb
2280 CMa	06 44.8	−27 38	11.8	5.6×3.2	Sb
2442 Vol	07 36.4	−69 32	11.2	6.0×5.5	SBb
LMC Dor	05 23.6	−69 45	0.1	650×550	Large Magellanic Cloud. Contains 30 Doradus and three planetary nebulae, NGC 1714, 1722 and 1743.

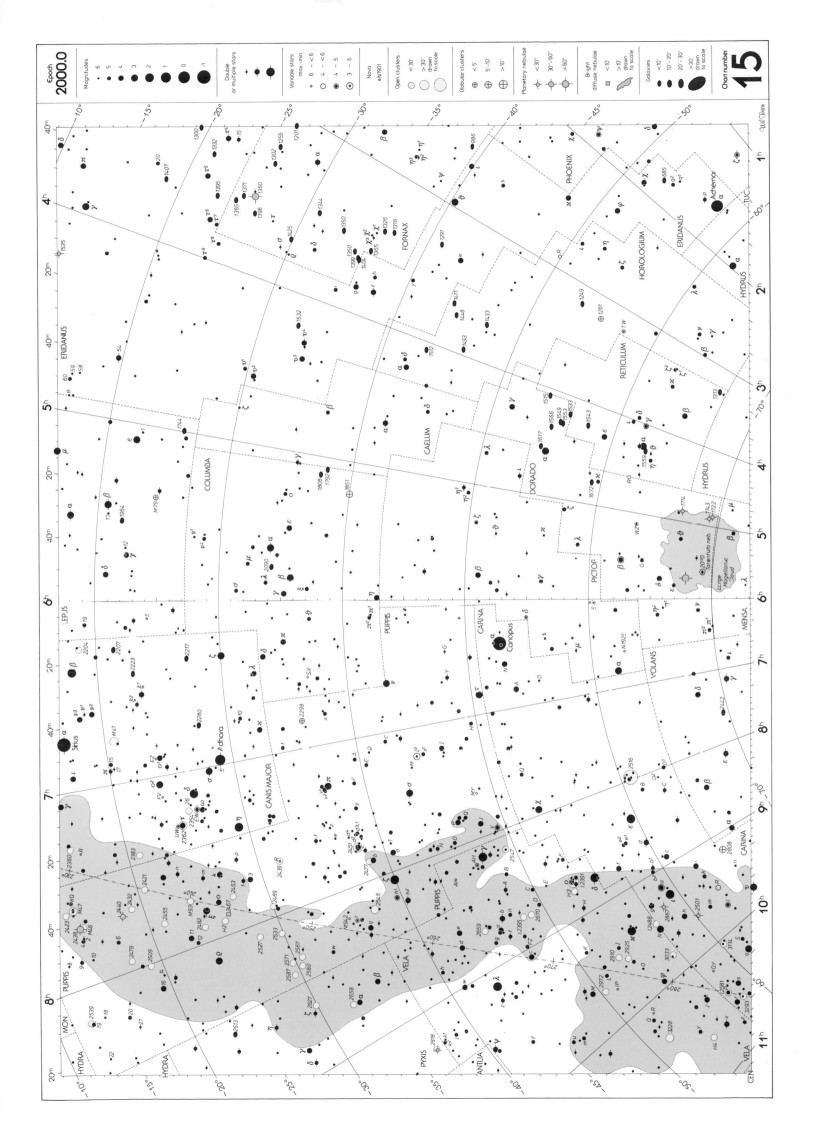

Chart 16. RA 8h to 12h. Dec −20° to −70°

Variable stars

	RA h m	Dec °	Range	Type	Period d	Spectrum
S Ant	09 32.3	−28 38	6.4– 6.9	EW	0.65	A
U Ant	10 35.2	−39 34	5.7– 6.8	Irr	—	N
η Car	10 45.1	−59 41	−0.8– 7.9	Irr	—	Pec
R Car	09 32.2	−62 47	3.9–10.5	M	308.7	M
U Car	10 57.8	−59 44	5.7– 7.0	Cep	38.77	F–G
ZZ (1) Car	09 45.2	−62 30	3.3– 4.2	Cep	35.53	F–K
S Pyx	09 05.1	−23 05	8.0–14.2	M	206.4	M
Z Vel	09 52.9	−54 11	7.8–14.8	M	421.6	M
AH Vel	08 12.0	−46 29	5.5– 5.9	Cep	4.23	F
AI Vel	08 14.1	−44 34	6.4– 7.1	δ Sct	0.11	A–F

Double stars

	RA h m	Dec °	PA	Sep	Mag	
δ Ant	10 29.6	−30 36	226	11.0	5.6, 9.6	
θ Ant	09 44.2	−27 46	005	0.1	5.4, 5.6	
π Cen	11 21.0	−54 29	128	0.4	4.3, 5.0	Binary, 39.2 y
β Hya	11 52.9	−33 54	008	0.9	4.7, 5.5	
44 Hya	10 34.0	−23 45	061	19.1	5.1,13.8	
δ Pyx	08 55.5	−27 41	AB 268	23.8	4.9,14.0	
			CD 017	2.5	11.0,11.0	
ε Pyx	09 09.9	−30 22	A+BC 147	17.8	5.6,10.5	
			BC 088	0.3	10.5,10.8	
			AD 340	35.4	5.6,13.5	
ζ Pyx	08 39.7	−29 34	061	52.4	4.9, 9.1	
η Pyx	08 37.9	−26 15	097	16.0	5.3,13.1	
ϰ Pyx	09 08.0	−25 52	263	2.1	4.6, 9.8	
γ Vel	08 09.5	−47 20	AB 220	41.2	1.9, 4.2	
			AC 151	62.3	8.2	
			AD 141	93.5	9.1	
			DE 146	1.8	12.5	
δ Vel	08 44.7	−54 43	AB 153	2.6	2.1, 5.1	
			AC 061	69.2	11.0	
			CD 102	6.2	13.5	
μ Vel	10 46.8	−49 25	055	2.3	2.7, 6.4	Binary, 116 y
b Vel	08 40.6	−46 39	058	37.5	3.8,10.2	

Open clusters

NGC	RA	Dec	Diam	Mag	N*	
2516 Car	07 58.3	−60 52	30	3.8	80	
2527 Pup	08 05.3	−28 10	22	6.5	40	
2533 Pup	08 07.0	−29 54	3.5	7.6	60	
2546 Pup	08 12.4	−37 38	41	6.3	40	
2547 Vel	08 10.7	−49 16	20	4.7	80	
2567 Pup	08 18.6	−30 38	10	7.4	40	
2571 Pup	08 18.9	−29 44	13	7.0	30	
2580 Pup	08 21.6	−30 19	8	9.7	50	
2587 Pup	08 23.5	−29 30	9	9.2	40	
2627 Pyx	08 37.3	−29 57	11	8.4	60	
2658 Pyx	08 43.4	−32 39	12	9.2	80	
2669 Vel	08 44.9	−52 58	12	6.1	40	
2670 Vel	08 45.5	−48 47	9	7.8	30	
2910 Vel	09 30.4	−52 54	5	7.2	30	
2925 Vel	09 33.7	−53 26	12	8.3	40	
2972 Vel	09 40.3	−50 20	4	9.9	25	
3033 Vel	09 48.8	−56 25	5	8.8	50	
3114 Car	10 02.7	−60 07	35	4.2	—	
3228 Vel	10 21.8	−51 43	18	6.0	15	
3532 Car	11 06.4	−58 40	55	3.0	150	
3572 Car	11 10.4	−60 14	7	6.6	35	
3590 Car	11 12.9	−60 47	4	8.2	25	
3680 Car	11 25.7	−43 15	12	7.6	30	
3766 Cen	11 36.1	−61 37	12	5.3	100	
3960 Cen	11 50.9	−55 42	7	8.3	45	
IC 2391 Vel	08 40.2	−53 04	50	2.5	30	o Velorum cluster
IC 2395 Vel	08 41.1	−48 12	8	4.6	40	
IC 2488 Vel	09 27.6	−56 59	15	7.4	70	
IC 2581 Car	10 27.4	−57 38	8	4.3	25	
IC 2602 Car	10 43.2	−64 24	50	1.9	60	θ Carinae cluster
IC 2714 Car	11 17.9	−62 42	12	8.2	100	
IC 2944 Cen	11 36.6	−63 02	15	4.5	30	λ Centauri cluster
Mel 101 Car	10 42.1	−65 06	14	8.0	50	
Mel 105 Car	11 19.5	−63 30	4	8.5	70	

Globular clusters

NGC	RA	Dec	Diam	Mag	
2808 Car	09 12.0	−64 52	13.8	6.3	
3201 Vel	10 17.6	−46 25	18.2	6.7	Dun 445

Planetary nebulae

NGC	RA	Dec	Diam	Mag	Mag*
2818 Pyx	09 16.0	−36 28	38	13.0	13.0
2867 Car	09 21.4	−58 19	11	9.7	13.6
3132 Vel	10 07.7	−40 26	47	8.2	10.1
3211 Car	10 17.8	−62 40	12	11.8	—
3918 Cen	11 50.3	−57 11	12	8.4	10.8
IC 2448 Car	09 07.1	−69 57	8	11.5	12.9
IC 2501 Car	09 38.8	−60 05	25	11.3	—
IC 2621 Car	11 00.3	−65 15	5	—	13.6

Galaxies

NGC	RA	Dec	Diam	Mag	Type
2613 Pyx	08 33.4	−22 58	7.2×2.1	10.4	Sb
2784 Hya	09 12.3	−24 10	5.1×2.3	10.1	SO
2835 Hya	09 17.9	−22 21	6.3×4.4	11.1	Sp
2997 Ant	09 45.6	−31 11	8.1×6.5	10.6	Sc
3109 Hya	10 03.1	−26 09	14.5×3.5	10.4	Irr
3223 Ant	10 21.6	−34 16	4.1×2.6	11.8	Sb
3347 Ant	10 42.8	−36 22	4.4×2.6	12.5	SBb
3511 Crt	11 03.4	−23 05	5.4×2.2	11.6	Sc
3513 Crt	11 03.8	−23 15	2.8×2.3	12.0	SBc
3557 Cen	11 10.0	−37 32	4.0×2.7	10.4	E3
3585 Hya	11 13.3	−26 45	2.9×1.6	10.4	E5
3621 Hya	11 18.3	−32 49	10.0×6.5	10.0	Sc
3923 Hya	11 51.0	−28 48	2.9×1.9	10.1	E3

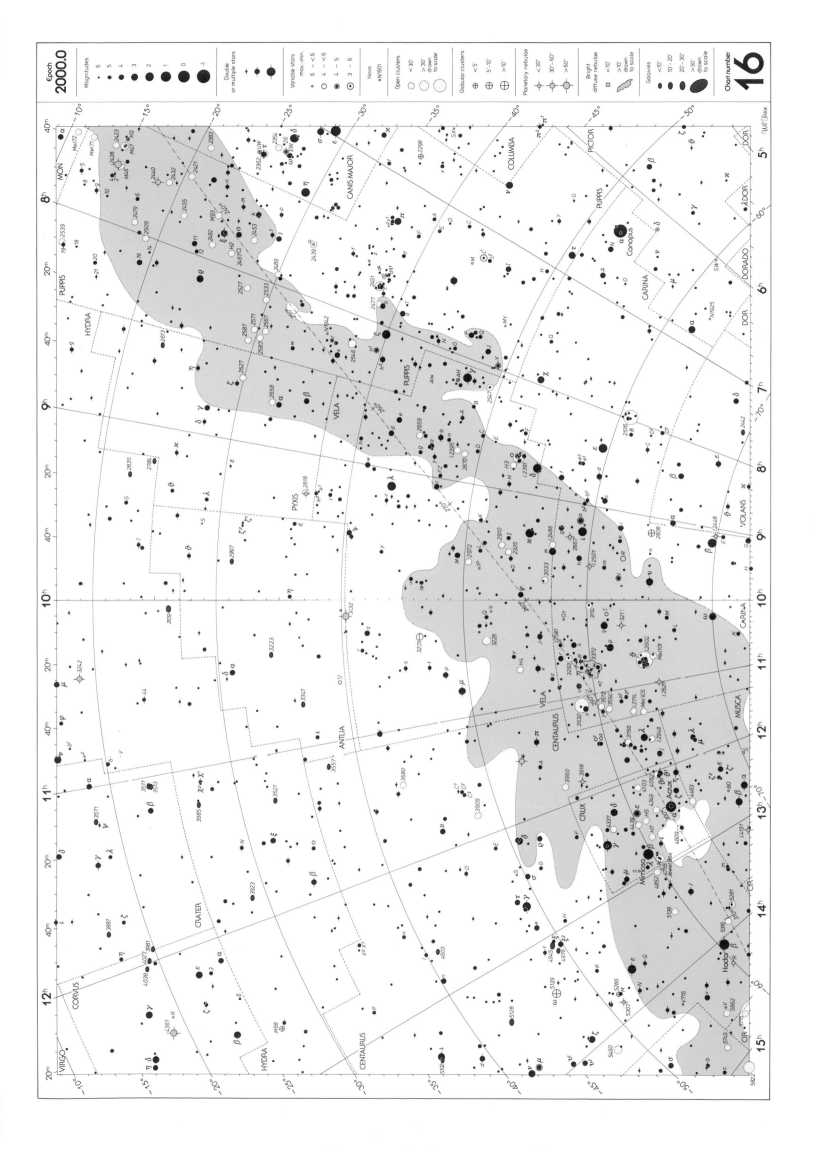

Chart 17. RA 12h to 16h. Dec −20° to −70°

Variable stars

Star		RA h m	Dec °	Range	Type	Period d	Spectrum
μ	Cen	13 49.6	−42 28	2.9– 3.5	Irr	–	B
R	Cen	14 16.6	−59 55	5.3–11.8	M	546.2	M
S	Cen	12 24.6	−49 26	6.0– 7.0	SR	65	N
T	Cen	13 41.8	−33 36	5.5– 9.0	SR	90	K–M
V	Cen	14 32.5	−56 53	6.4– 7.2	Cep	5.49	F–G
W	Cen	11 55.0	−59 15	7.6–13.7	M	201.6	M
S	Cru	12 54.4	−58 26	6.2– 6.9	Cep	4.69	F–G
R	Hya	13 29.7	−23 17	4.0–10.0	M	389.6	M
GG	Lup	15 18.9	−40 47	5.4– 6.0	EB	2.16	B+A
R	Mus	12 42.1	−69 24	5.9– 6.7	Cep	7.48	F
R	Nor	15 36.0	−49 30	6.5–13.9	M	492.7	M
T	Nor	15 44.1	−54 59	6.2–13.6	M	242.6	M

Double stars

Star		RA	Dec	PA	Sep	Mag
α	Cen	14 39.6	−60 50	215	19.7	0.0, 1.2 Binary, 79.9 y
β	Cen	14 03.8	−60 22	251	1.3	0.7, 3.9
γ	Cen	12 41.5	−48 58	353	1.0	2.9, 2.9 Binary, 84.5 y
ε	Cen	13 39.9	−53 28	158	36.0	2.3,12.7
η	Cen	14 35.5	−42 09	270	5.0	2.6,13.5
3	Cen	13 51.8	−33 00	108	7.9	4.5, 6.0
4	Cen	13 53.2	−31 56	185	14.9	4.8, 8.4
α	Cir	14 42.5	−64 59	232	15.7	3.2, 8.6
γ	Cir	15 23.4	−59 19	033	0.6	5.1, 5.5 Binary, 180 y
δ	Cir	15 16.9	−60 57	270	50.0	5.1,13.4
ζ	Crv	12 20.6	−22 13	066	11.2	5.2,13.6
α	Cru	12 26.6	−63 06	{AB 115	4.4	1.4, 1.9
				{AC 202	90.1	1.0, 4.9
γ	Cru	12 31.2	−57 07	{AB 031	110.6	1.6, 6.7
				{AC 082	155.2	9.5
η	Cru	12 06.9	−64 37	299	44.0	4.2,11.7
θ¹	Cru	12 03.0	−63 19	325	4.5	4.3,13.6
ι	Cru	12 45.6	−60 59	022	26.9	4.7, 9.5
μ	Cru	12 54.6	−57 11	017	34.9	4.0, 5.2
R	Hya	13 29.7	−23 17	324	21.2	var,12.0
52	Hya	14 28.2	−29 30	{AB 130	0.1	5.8, 5.8
				{AB+C 279	4.2	10.0
				{AB+D 282	140.8	12.0
54	Hya	14 46.0	−15 27	126	8.6	5.1, 7.1
59	Hya	14 58.7	−27 39	335	0.8	6.3, 6.6
κ	Lup	15 11.9	−48 44	144	26.8	3.9, 5.8
μ	Lup	15 18.5	−47 53	{AB 142	1.2	5.1, 5.2
				{AC 130	23.7	7.2
ξ	Lup	15 56.9	−33 58	049	10.4	5.3, 5.8
π	Lup	15 05.1	−47 03	073	1.4	4.6, 4.7
τ¹	Lup	14 26.1	−45 13	204	148.2	4.6, 9.3
υ	Lup	14 24.7	−39 43	038	1.4	5.4,10.9
α	Mus	12 37.2	−69 08	316	29.6	2.7,12.8
β	Mus	12 46.3	−68 06	014	1.4	3.7, 4.0
ζ²	Mus	12 22.1	−67 31	130	32.4	5.2,10.6
θ	Mus	13 08.1	−65 18	187	5.3	5.7, 7.3
π	Sco	15 58.9	−26 07	132	50.4	2.9,12.1
2	Sco	15 53.6	−25 20	274	2.5	4.7, 7.4

Open clusters

NGC		RA	Dec	Diam	Mag	N*
4103	Cru	12 06.7	−61 15	7	7.4	45
4337	Cru	12 23.9	−58 08	3.5	8.9	—
4439	Cru	12 28.4	−60 06	4	8.4	—
4349	Cru	12 24.5	−61 54	16	7.4	30
4463	Mus	12 30.0	−64 48	5	7.2	30
4609	Cru	12 42.3	−62 58	5	6.9	40
4755	Cru	12 53.6	−60 20	10	4.2	50+ Jewel Box: κ Crucis
5138	Cen	13 27.3	−59 01	8	7.6	40
5281	Cen	13 46.6	−62 54	5	5.9	40
5316	Cen	13 53.9	−61 52	14	6.0	80
5460	Cen	14 07.6	−48 19	25	5.6	40
5617	Cen	14 29.8	−60 43	10	6.3	80
5662	Cen	14 35.2	−56 33	12	5.5	70
5749	Lup	14 48.9	−54 31	8	8.8	30
5822	Lup	15 05.2	−54 21	40	6.5	150
5823	Cir	15 05.7	−55 36	10	7.9	100
5925	Nor	15 27.7	−54 31	15	8.4	120
5999	Nor	15 52.2	−56 28	5	9.0	40
H.5	Cru	12 29.0	−60 46	6	7.1	—

Globular clusters

NGC		RA	Dec	Diam	Mag	
4590	Hya	12 39.5	−26 45	12.0	8.2	M68
5139	Cen	13 26.8	−47 29	36.3	3.6	ω Centauri
5286	Cen	13 46.4	−51 22	9.1	7.6	
5824	Lup	15 04.0	−33 04	6.2	9.0	
5897	Lib	15 17.4	−21 01	12.6	8.6	
5927	Lup	15 28.0	−50 40	12.0	8.3	H IV 19
5986	Lup	15 46.1	−37 47	9.8	7.1	Dun 552

Planetary nebulae

NGC		RA	Dec	Diam	Mag	Mag*
5882	Lup	15 16.8	−45 39	7	10.5	12.0
IC 4191	Mus	13 08.8	−67 39	5	12.0	—
IC 4406	Lup	14 22.4	−44 09	28	10.6	14.7

Nebulae

NGC		RA	Dec	Diam	Mag*	
5367	Cen	13 57.7	−39 59	4× 3		Includes IC 4347. Double nucleus
—	Cru	12 53	−63	400×300		Coal Sack: dark nebula, area 26.2 sq deg

Galaxies

NGC		RA	Dec	Diam	Mag	Type
4603	Cen	12 40.9	−40 59	3.8× 2.5	12.0	Sc
4945	Cen	13 05.4	−49 28	20.0× 4.4	9.5	SBc
4976	Cen	13 08.6	−49 30	4.3× 2.6	10.2	E4p
5078	Hya	13 19.8	−27 24	3.2× 1.7	12.0	Sa
5101	Hya	13 21.8	−27 26	5.5× 4.9	11.7	SBa
5061	Hya	13 18.1	−26 50	2.6× 2.3	11.7	E2
5068	Vir	13 18.9	−21 02	6.9× 6.3	10.8	SBc
5085	Hya	13 20.3	−24 26	3.4× 3.0	10.8	Sb
5102	Cen	13 22.0	−36 38	9.3× 3.5	11.9	SO
5128	Cen	13 25.5	−43 01	18.2×14.3	9.6	SOp Centaurus A
5236	Hya	13 37.0	−29 52	11.2×10.2	7.0	Sc M83
5253	Cen	13 39.9	−31 39	4.0× 1.7	8.2	E5
5483	Cen	14 10.4	−43 19	3.1× 2.8	10.6	Sc
5643	Lup	14 32.7	−44 10	4.6× 4.1	10.7	SBO
IC 4296	Cen	13 36.6	−33 58	—	10.6	EO

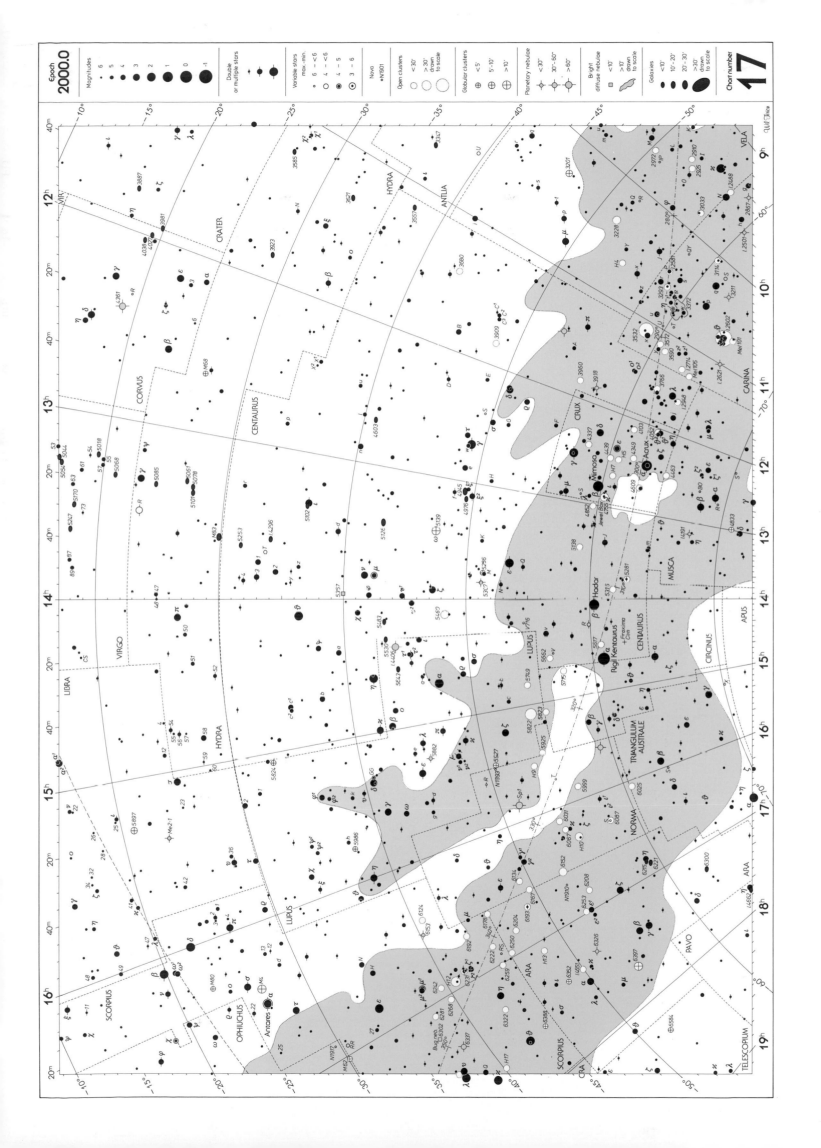

Chart 18. RA 16h to 20h. Dec −20° to −70°

Variable stars

	RA h m	Dec ° ′	Range	Type	Period d	Spectrum
ε CrA	18 58.7	−37 06	4.7– 5.0	EW	0.59	F
S Nor	16 18.9	−57 54	6.1– 6.8	Cep	9.75	F-G
κ Pav	18 56.9	−67 14	3.9– 4.7	Cep.W	9.09	F
λ Pav	18 52.2	−62 11	3.4– 4.3	Irr	–	B
S Pav	19 55.2	−59 12	6.6–10.4	SR	386.0	M
BM Sco	17 41.0	−32 13	6.8– 8.7	SR	850.0	K
RR Sco	16 55.6	−30 35	5.0–12.4	M	279.4	M
RS Sco	16 56.6	−45 06	6.2–13.0	M	320.0	M
RU Sco	17 42.4	−43 45	7.8–13.7	M	369.2	M
RR Sgr	19 55.9	−29 11	5.6–14.0	M	334.6	M
RU Sgr	19 58.7	−41 51	6.0–13.8	M	240.3	M
RY Sgr	19 16.5	−33 31	6.0–15.0	RCrB	–	Gp
W Sgr	18 05.0	−29 35	4.3– 5.1	Cep	7.59	F-G
S TrA	16 01.2	−63 47	6.1– 6.8	Cep	6.32	F

Double stars

	RA h m	Dec ° ′	PA	Sep	Mags
γ Ara	17 25.4	−56 23	AB 328 / AC 066	17.9 / 41.6	3.3,10.3 / 4.8, 5.1
γ CrA	19 06.4	−37 04	109	1.3	4.8, 5.1 Binary, 120.4 y
λ CrA	18 43.8	−38 19	214	29.2	5.1, 9.7
ε Nor	16 27.0	−47 03	335	22.8	4.8, 7.5
ι¹ Nor	16 03.5	−57 47	100	0.2	5.3, 5.5 Binary, 26.9 y
36 Oph	17 15.3	−26 36	150	4.7	5.1, 5.1 Binary, 549 y
ξ Pav	18 23.2	−61 30	154	3.3	4.4, 8.6
α Sco	16 29.4	−26 26	273	2.7	1.2, 5.4 Antares, Binary, 878 y
σ Sco	16 21.2	−25 36	273	20.0	2.9, 8.5
12 Sco	16 12.3	−28 25	073	4.0	5.9, 7.9
β¹ Sgr	19 22.6	−44 28	077	28.3	3.9, 8.0 Wide naked-eye pair with β² (4.3y)
σ Sgr	19 02.6	−29 53	320	0.3	3.3, 3.4
η Sgr	18 17.6	−36 46	105	3.6	3.2, 7.8
π Sgr	19 09.8	−21 01	AB 150 / AB+C 122	0.1 / 0.4	3.7, 3.7 / 5.9
21 Sgr	18 25.3	−20 32	289	1.8	4.9, 7.4
52 Sgr	19 36.7	−24 53	170	2.5	4.7, 9.2

Open clusters

NGC	RA	Dec	Diam	Mag	N*	
5050 TrA	16 03.7	−60 30	12	5.1	60	
6067 Nor	16 13.2	−54 13	13	5.6	100	
6087 Nor	16 18.9	−57 54	12	5.4	40	S Normae cluster
6124 Sco	16 25.6	−40 40	29	5.8	100	
6134 Nor	16 27.7	−49 09	7	7.2	–	
6167 Nor	16 34.4	−49 36	8	6.7	–	
6178 Sco	16 35.7	−45 38	4	7.2	12	
6193 Ara	16 41.3	−48 46	15	5.2	30	
6204 Ara	16 46.5	−47 01	5	8.2	45	
6208 Ara	16 49.5	−53 49	16	7.2	60	
6231 Sco	16 54.0	−41 48	15	2.6	–	
6242 Sco	16 55.6	−39 30	9	6.4	–	
6250 Ara	16 58.0	−45 48	8	5.9	60	
6281 Sco	17 04.8	−37 54	8	5.4	40	Nebulosity
6383 Sco	17 34.8	−32 34	5	5.5	80	
6405 Sco	17 40.1	−32 13	15	4.2	40	M6 Butterfly cluster
6416 Sco	17 44.4	−32 21	18	5.7	40	
6520 Sco	18 03.4	−27 54	6	7.6	60	In M20 (Trifid)
6531 Sgr	18 04.6	−22 30	13	5.9	70	M21
6546 Sgr	18 07.2	−23 20	13	8.0	150	
6475 Sco	17 53.9	−34 49	80	3.3	80	M7
IC 4651 Ara	17 24.7	−49 57	12	6.9	80	
H.10 Nor	16 19.9	−54 59	30	–	30	
H.13 Ara	17 05.4	−48 11	15	–	15	

Globular clusters

NGC	RA	Dec	Diam	Mag	
6093 Sco	16 17.0	−22 59	8.9	7.2	M80
6121 Nor	16 23.6	−26 32	26.3	5.9	M4
6266 Oph	17 01.2	−30 07	14.1	6.6	M62
6273 Oph	17 02.6	−26 16	13.5	7.1	M19
6304 Oph	17 14.5	−29 28	6.8	8.4	
6352 Ara	17 25.5	−48 25	7.1	8.1	
6355 Oph	17 24.0	−26 21	5.0	9.6	
6362 Ara	17 31.9	−67 03	10.7	8.3	
6388 Sco	17 36.3	−44 44	8.7	6.8	
6397 Ara	17 40.7	−53 40	25.7	5.6	Dun 473
6541 CrA	18 08.0	−43 42	13.1	6.6	
6544 Sgr	18 07.3	−25 00	8.9	8.2	
6553 Sgr	18 09.3	−25 54	8.1	8.2	
6558 Sgr	18 10.3	−31 46	3.7	–	
6569 Sgr	18 13.6	−31 50	5.8	8.7	H150
6624 Sgr	18 23.7	−30 22	5.9	8.3	M28
6626 Sgr	18 24.5	−24 52	11.2	6.9	M69
6637 Sgr	18 31.4	−32 21	7.1	7.7	H151
6638 Sgr	18 30.9	−25 30	5.0	9.2	M22
6652 Sgr	18 35.8	−32 59	3.5	8.9	M54
6656 Sgr	18 36.4	−23 54	24.0	5.1	M70
6715 Sgr	18 55.1	−30 09	9.1	7.7	M55
6752 Pav	19 10.9	−59 59	20.4	5.4	
6681 Sgr	18 43.2	−32 18	7.8	8.1	
6809 Sgr	19 40.0	−30 58	19.0	6.9	

Planetary nebulae

NGC	RA	Dec	Diam	Mag	Mag*	
6153 Sco	16 31.5	−40 15	25	11.5	–	
6302 Sco	17 13.7	−37 06	50	12.8	–	Bug Nebula
6337 Sco	17 22.3	−38 29	48	–	14.7	
6629 Sgr	18 25.7	−23 12	15	–	12.8	
6644 Sgr	18 32.6	−25 08	3	12.2	15.9	
IC 1297 CrA	19 17.4	−39 37	7	–	12.9v	RU CrA
IC 4699 Tel	18 18.5	−45 59	10	11.9	–	

Nebulae

NGC	RA	Dec	Diam	Mag	
6514 Sgr	18 02.6	−23 02	29×27	7.6	Trifid Nebula. M20
6523 Sgr	18 08.3	−24 23	90×40	–	Lagoon Nebula, M8

Galaxies

NGC	RA	Dec	Mag	Diam	Type
6215 Ara	16 51.1	−58 59	11.8	2.0× 1.6	Sc
6221 Ara	16 52.8	−59 13	11.5	3.2× 2.3	SBc
6684 Pav	18 49.0	−65 11	10.4	3.7× 2.7	SB0
6744 Pav	19 09.8	−63 51	9.0	15.5×10.2	SBb
6753 Pav	19 11.4	−57 03	11.9	2.5× 2.2	Sb
IC 4662 Pav	17 47.1	−64 38	11.4	2.2× 1.4	Irr

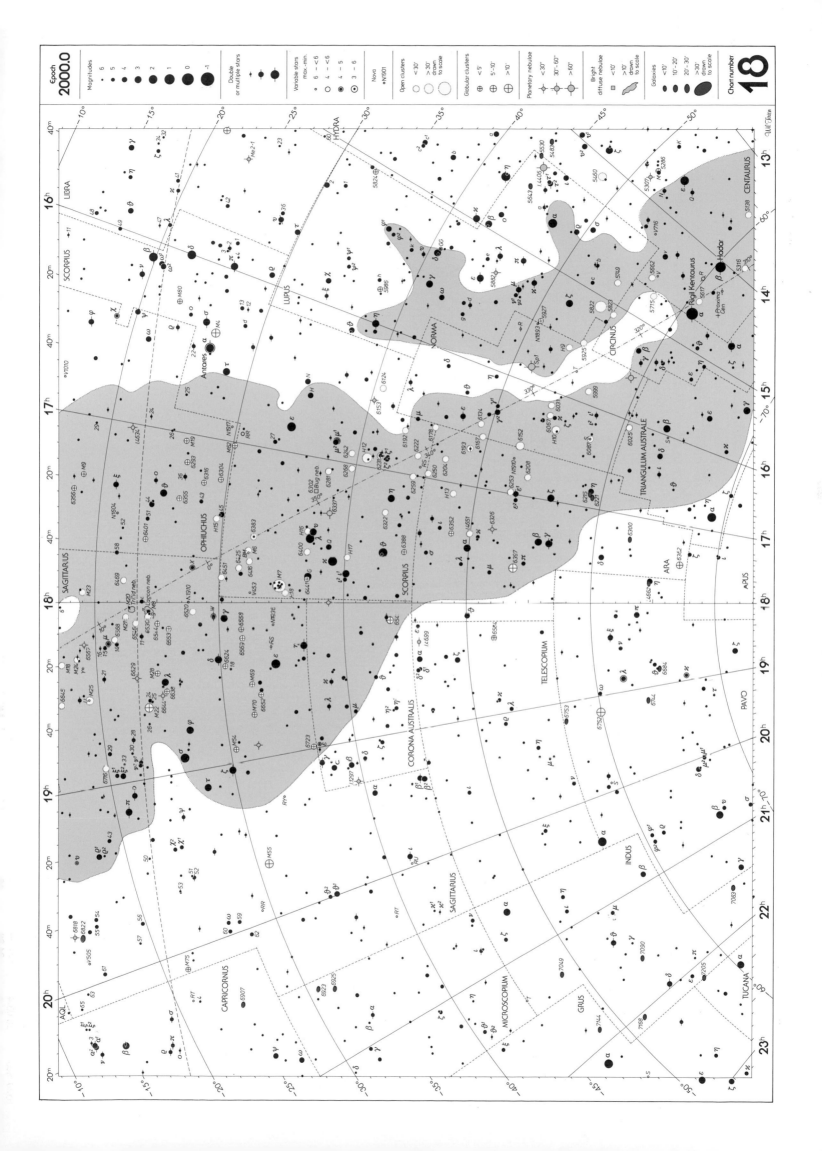

Chart 19. RA 20h to 0h. Dec −20° to −70°

Variable stars

	RA h m	Dec ° ′	Range	Type	Period d	Spectrum
RT Cap	20 17.1	−21 19	8.9–11.7 pho.	SR	393	M
S Gru	22 26.1	−48 26	6.0–15.0	M	401.4	M
T Gru	22 25.7	−37 34	7.8–12.3	M	136.5	M
T Mic	20 27.9	−28 16	7.7– 9.6	SR	344	M
Y Pav	21 24.3	−69 44	5.7– 8.5	SR	233	N
SX Pav	21 28.7	−69 30	5.4– 6.0	SR	50	M
RT Sgr	20 17.7	−39 07	6.0–14.1	M	305.3	M

Double stars

	RA	Dec	PA	Sep	Mag	
41 Aqr	22 14.3	−21 04	{AB 114 / AC 043	5.0 / 212.1	5.6, 7.1 / 9.0	
86 Aqr	23 06.7	−23 45	083	2.9	4.5,14.5	
89 Aqr	23 09.9	−22 27	007	0.4	5.1, 5.9	
ζ Cap	21 26.7	−22 25	013	21.3	3.7,12.3	
24 Cap	21 07.1	−25 00	186	26.2	4.6,11.7	
δ Ind	21 57.9	−55 00	323	0.1	5.3, 5.3	Binary, 12.0 y
θ Ind	21 19.9	−53 27	275	6.0	4.5, 7.0	
θ Gru	23 06.9	−43 31	075	1.1	4.5, 7.0	
υ Gru	23 06.9	−38 54	211	1.1	5.7, 8.0	
α Mic	20 50.0	−33 47	166	20.5	5.0,10.0	
θ² Mic	21 24.4	−41 00	{AB 267 / AC 066	0.5 / 78.4	6.4, 7.0 / 10.5	
β PsA	22 31.5	−32 21	172	30.4	4.4, 7.9	Optical
γ PsA	22 52.5	−32 53	262	4.2	4.5, 8.0	
δ PsA	22 55.9	−32 32	244	5.0	4.2, 9.2	
η PsA	22 00.8	−28 27	115	1.7	5.8, 6.8	
6 PsA	21 32.2	−33 57	059	6.8	6.0,13.3	
8 PsA	21 36.2	−26 10	008	18.4	5.7,13.9	
δ Scl	23 48.9	−28 08	{AB 243 / AC 297	3.3 / 74.3	4.5,11.5 / 9.3	
χ² Sgr	20 23.9	−42 25	234	0.8	6.0, 6.9	
δ Tuc	22 27.3	−64 58	282	6.9	4.5, 9.0	

Globular clusters

NGC	RA	Dec	Diam	Mag	
6864 Sgr	20 06.1	−21 55	6	8.6	M75
7099 Cap	21 40.4	−23 11	11	7.5	M30

Planetary nebula

NGC	RA	Dec	Diam	Mag	Mag*
7293 Aqr	22 29.6	−20 48	770	6.5	13.5 Helix Nebula

Galaxies

NGC	RA	Dec	Mag	Diam	Type
6907 Cap	20 25.1	−24 49	11.3	3.4×3.0	SBb
6923 Mic	20 31.7	−30 50	12.1	2.5×1.4	Sb
6925 Mic	20 34.3	−31 59	11.3	4.1×1.6	Sb
7049 Ind	21 19.0	−48 34	10.7	2.8×2.2	SO
7083 Ind	21 35.7	−63 54	11.8	4.5×2.9	Sb
7090 Ind	21 36.5	−54 33	11.1	7.1×1.4	SBc
7144 Gru	21 52.7	−48 15	10.7	2.5×2.3	EO
7168 Ind	22 02.1	−51 45	12.6	2.0×1.6	E3
7172 PsA	22 02.0	−31 52	11.9	2.2×1.3	S
7174 PsA	22 02.1	−31 59	12.6	1.3×0.7	S
7184 Aqr	22 02.7	−20 49	12.0	5.8×1.8	Sb
7205 Ind	22 08.5	−57 25	11.4	4.3×2.2	Sb
7314 PsA	22 35.8	−26 03	10.9	4.6×2.3	Sc Arp 14
7410 Gru	22 55.0	−39 40	10.4	5.5×2.0	SBa
7412 Gru	22 55.8	−42 39	11.4	4.0×3.1	SBb
7418 Gru	22 56.6	−37 02	11.4	3.3×2.8	SBc
7424 Gru	22 57.3	−41 04	11.0	7.6×6.8	SBc
7456 Gru	23 02.1	−39 35	11.9	5.9×1.8	Sc
7496 Gru	23 09.8	−43 26	11.1	3.5×2.8	SBb
7531 Gru	23 14.8	−43 36	11.3	3.5×1.5	Sb
7552 Gru	23 16.2	−42 35	10.7	3.5×2.5	SBb
7582 Gru	23 18.4	−42 22	10.6	4.6×2.2	SBb
7599 Gru	23 19.3	−42 15	11.4	4.4×1.5	Sc
7713 Scl	23 36.5	−37 56	11.6	4.3×2.0	SBd
7755 Scl	23 47.9	−30 31	11.8	3.7×3.0	SBb
7793 Scl	23 57.8	−32 35	9.1	9.1×6.6	Sd
IC 1459 Gru	22 57.2	−36 28	10.0	–	E3
IC 5267 Gru	22 57.2	−43 24	10.5	5.0×4.1	SO
IC 5273 Gru	22 59.5	−37 42	11.4	2.9×2.1	SBc
IC 5332 Scl	23 34.5	−36 06	10.6	6.6×5.1	Sd

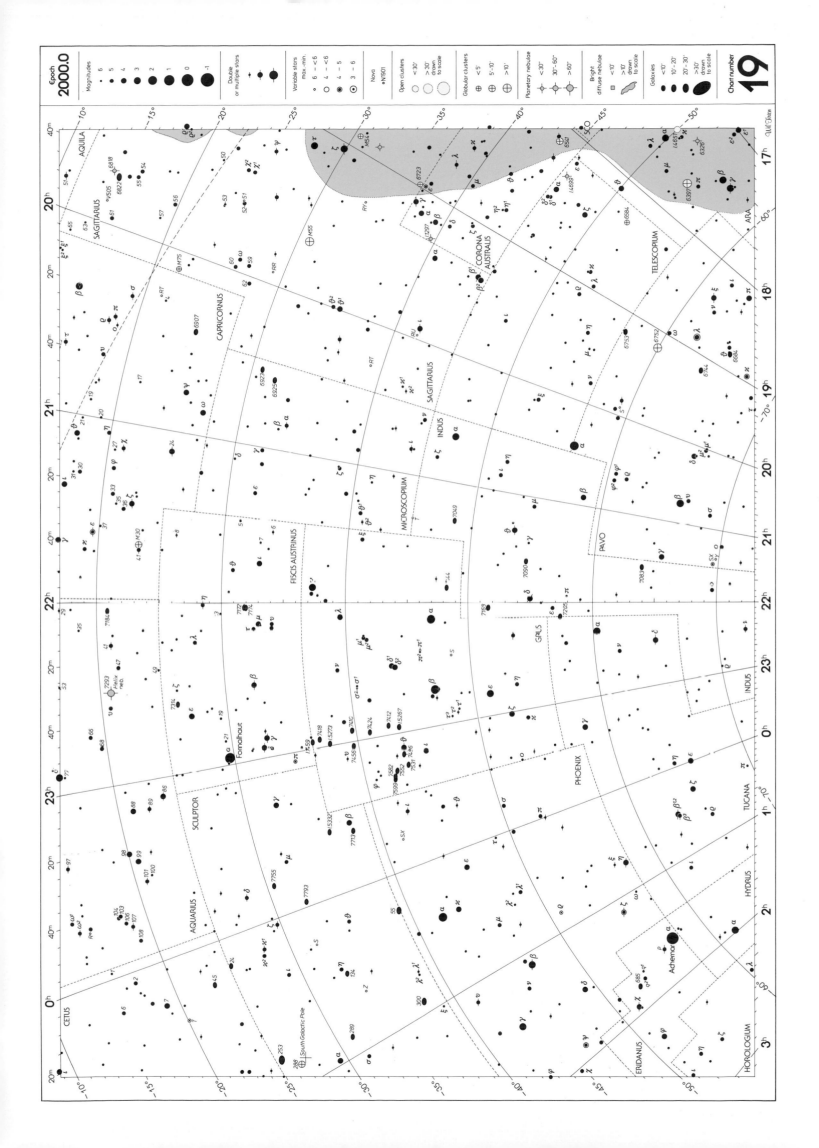

Chart 20. *Far south; declination below −70°*

Variable stars

		RA h m	Dec °	Type	Range	Period d	Spectrum
θ	Aps	14 05.3	−76 48	SR	6.4– 8.6	119	M
RS	Cha	08 43.2	−79 04	EA+δ Sct	6.0– 6.7	1.67	A–F
TZ	Men	05 30.2	−84 47	EA	6.2– 6.9	8.57	B
ε	Oct	22 20.0	−80 26	SR	4.9– 5.4	55	M
R	Oct	05 26.1	−86 23	M	6.4–13.2	405.6	M

Double stars

		RA	Dec	PA	Sep	Mag	
δ	Aps	16 20.3	−78 42	012	102.9	4.7, 5.1	
δ¹	Cha	10 45.3	−80 28	076	0.6	6.1, 6.4	
ε	Cha	11 59.6	−78 13	188	0.9	5.4, 6.0	
θ	Cha	08 20.6	−77 29	250	31.0	4.4,12.1	
τ²	Hyi	01 47.8	−80 11	028	39.7	6.1,13.5	
γ	Men	05 31.9	−76 20	107	38.2	5.2,11.2	
ι	Oct	12 55.0	−85 07	230	0.6	6.0, 6.5	
λ	Oct	21 50.9	−82 43	070	3.1	5.4, 7.7	
μ²	Oct	20 41.7	−75 21	017	17.4	7.1, 7.6	
γ²	Vol	07 08.8	−70 30	300	13.6	4.0, 5.9	
ζ	Vol	07 41.8	−72 36	116	16.7	4.0, 9.8	
θ	Vol	08 39.1	−70 23	AC 108	45.0	5.3,10.3	A is a close double
κ	Vol	08 19.8	−71 31	{AB 057 / BC 030}	65.0 / 37.7	5.4, 5.7 / 8.5	

Globular clusters

NGC	RA	Dec	Diam	Mag	
104 Tuc	00 24.1	−72 05	30.9	4.0	47 Tucanae
362 Tuc	01 03.2	−70 51	12.9	6.6	
4372 Mus	12 25.8	−72 40	18.6	7.8	
4833 Mus	12 59.6	−70 53	13.5	7.3	
6101 Aps	16 25.8	−72 12	10.7	9.3	
IC 4499 Aps	15 00.3	−82 13	7.6	10.6	

Galaxies

NGC	RA	Dec	Mag	Diam	Type
3059 Car	09 50.2	−73 55	11.8	3.2×3.0	SBb
5967 Aps	15 48.1	−75 40	12.5	2.9×1.8	SBc
SMC Tuc	00 52.7	−72 50	2.3	280×160	Small Cloud of Magellan

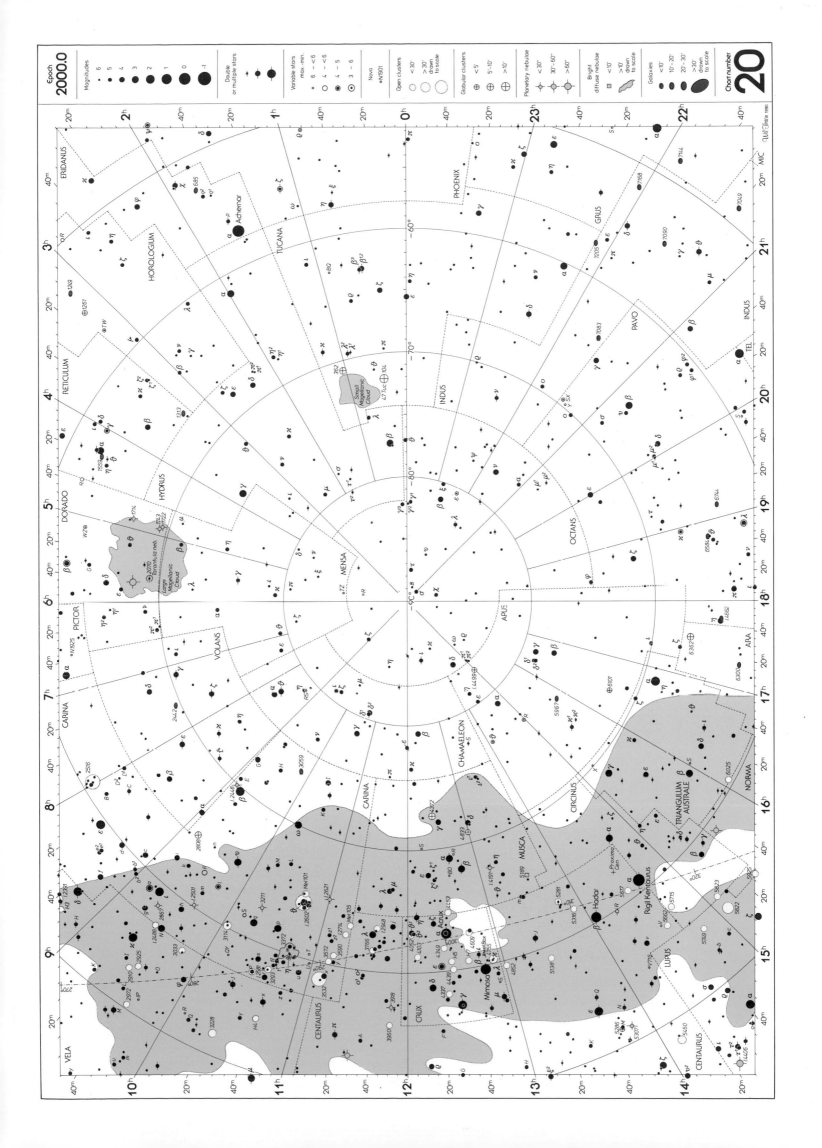

The all-sky maps

This final part of *Cambridge Star Atlas 2000.0* consists of six all-sky maps to show the general distribution of different objects in the sky. On each map you will find the northern and the southern hemispheres of the heavens, side by side. The maps are on an equal-area projection. That means that in spite of the inevitable distortion the actual area covered by one square degree remains the same, no matter where on the map it is measured. So the distribution, or density of objects, is not influenced by the map's projection.

The first map gives the positions of the constellations, against the usual grid of RA and Dec, as well as the position of the Galactic Equator, with tick marks for the galactic centre (0°) and 180° of galactic longitude. Stars down to magnitude 4.5 are plotted, plus some fainter stars to complete the constellation patterns, as on the monthly sky maps.

Distribution of open clusters

Here we see the open clusters plotted against the background of stars and constellations (in grey). As explained in the introduction to the main star charts the open clusters are found near the plane of the milky way, so on the map you will find most of these objects close to the Galactic Equator. All clusters brighter than 10.0 (which are also plotted on the charts) are shown as solid white discs. All fainter clusters, plotted on *Sky Atlas 2000.0* (see Sources and references) are plotted as open circles. This same rule is used for the globular clusters, and the planetary and difuse nebulae.

Distribution of globular clusters

This is quite different. The globular clusters form a huge halo around the milky way galaxy. So they are much more scattered than the open clusters. But since we are not in the centre of our galaxy (in fact, we are closer to the edge than to the centre) we see most of the globular clusters in the direction of the galactic heart (in Sagittarius). In the opposite direction (near 180° galactic longitude, in Auriga) they are almost absent.

Distribution of planetary nebulae

This lies somewhere in between the previous two cases. It is not limited to the spiral arms of the galaxy, nor do the planetary nebulae fill a halo, like the globular clusters. They fill a disc-like area, but much thicker than the milky way disc.

Distribution of diffuse nebulae

This brings us back to the disc of the galaxy. Like the open clusters, these are found along the spiral arms of the milky way.

Distribution of galaxies

This has no relation to our own galaxy. But, looking at the map, does not give that impression as galaxies are almost absent in the milky way area. This, however, has nothing to do with their real distribution in space. The enormous amounts of gas and dust in our galaxy blocks the light of most of the distant galaxies seen in that direction. So they look absent. The galaxies are obviously grouped into what astronomers call 'clusters' and 'superclusters'. On the map only the galaxies from the *Cambridge Star Atlas* are plotted, to avoid cluttering.

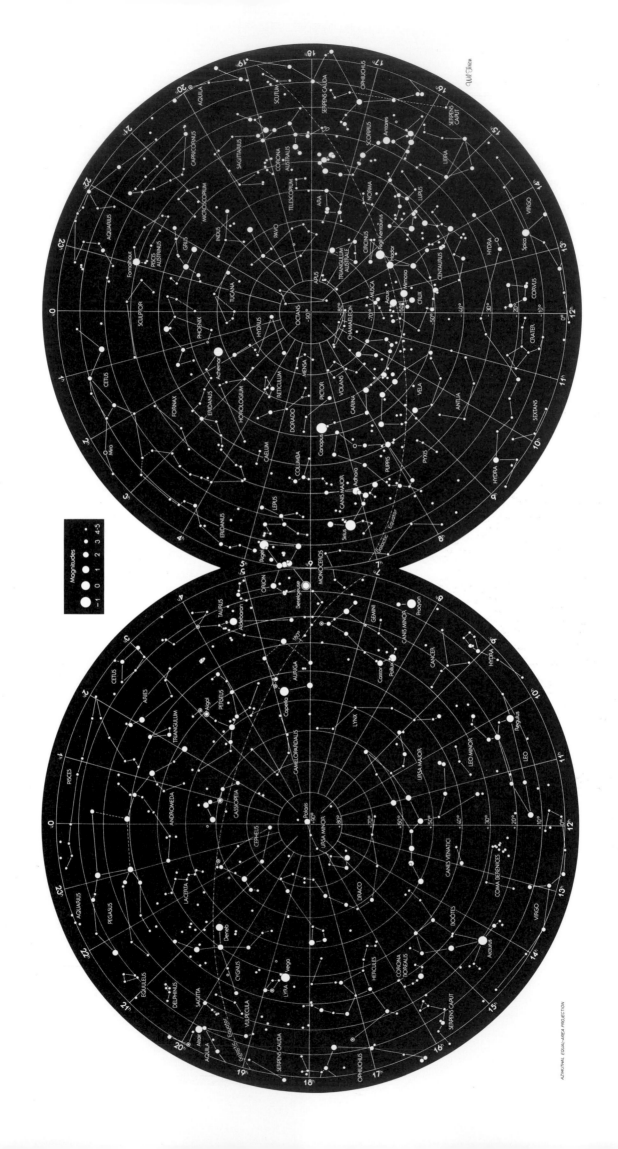

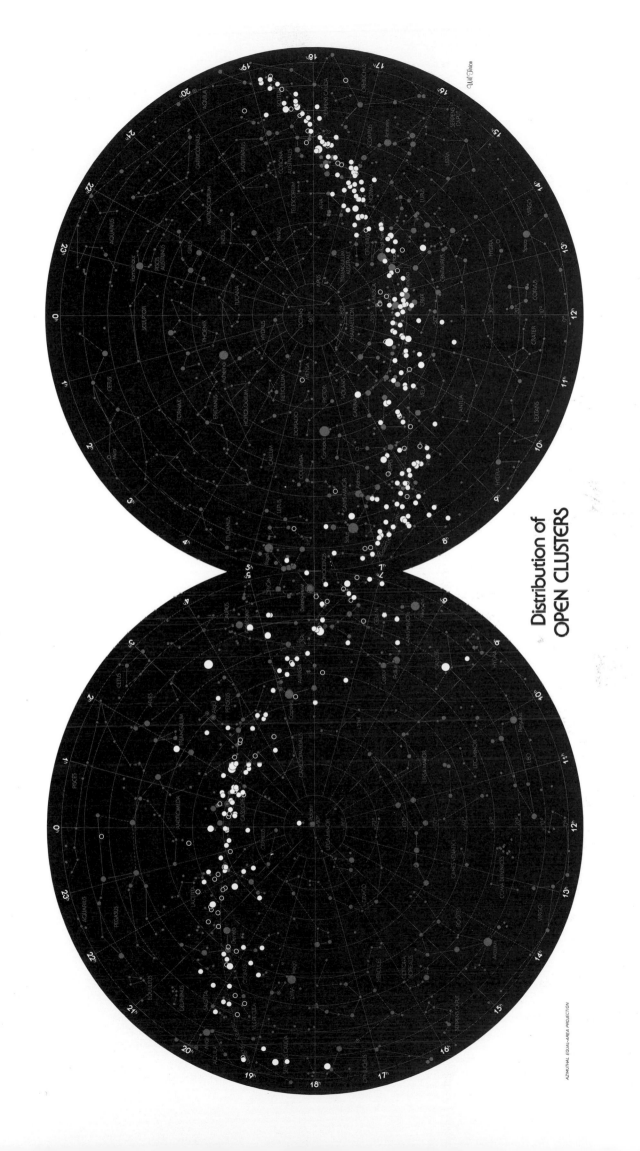

Distribution of
OPEN CLUSTERS

AZIMUTHAL EQUAL-AREA PROJECTION

Distribution of
GLOBULAR CLUSTERS

AZIMUTHAL EQUAL-AREA PROJECTION

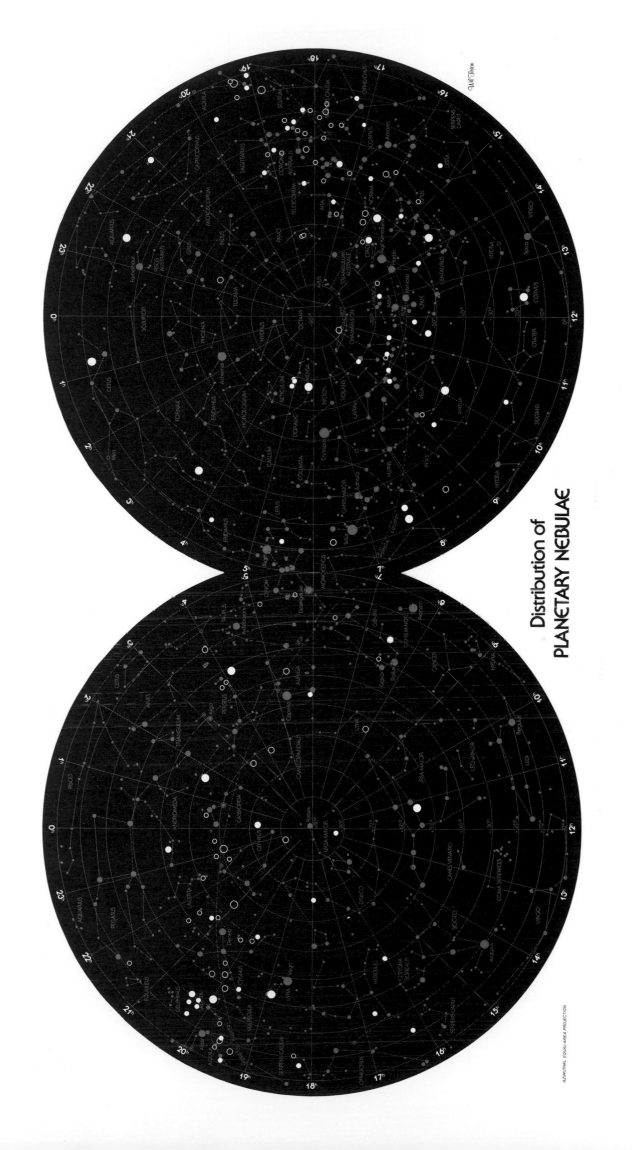

Distribution of
PLANETARY NEBULAE

Wil Tirion

AZIMUTHAL EQUAL-AREA PROJECTION

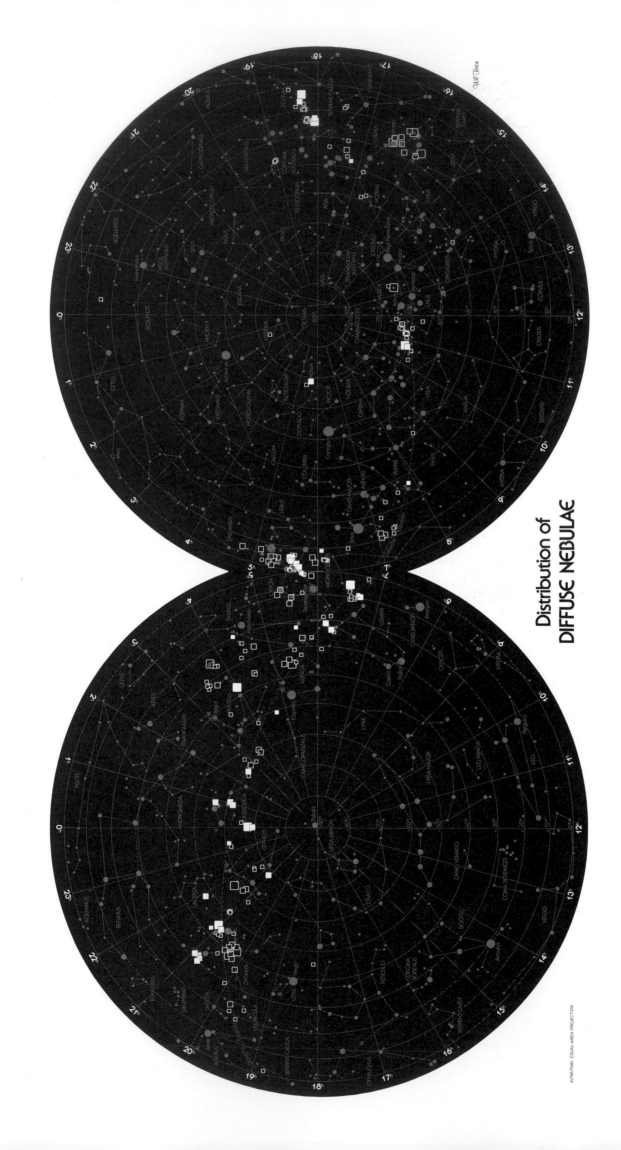

Distribution of
DIFFUSE NEBULAE

AZIMUTHAL EQUAL-AREA PROJECTION

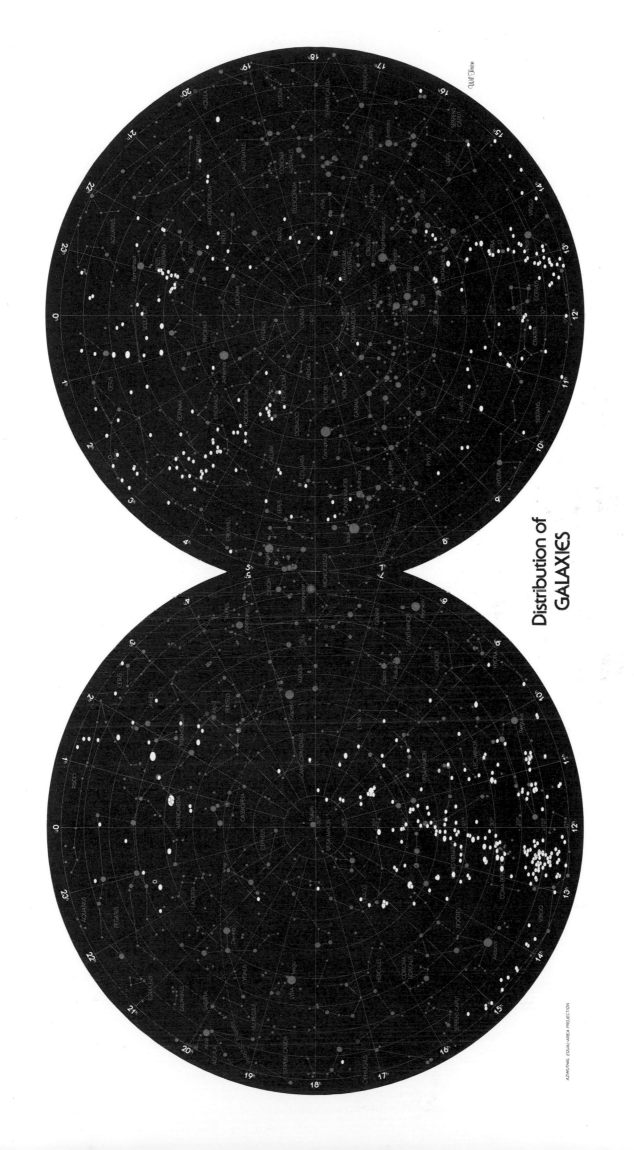

Distribution of
GALAXIES

Sources and references

Catalogues

Dorrit Hoffleit, *The Bright Star Catalogue*, 4th edn, Yale University Observatory, New Haven, Connecticut, 1982.

Smithsonian Astrophysical Observatory Star Catalog, Smithsonian Institution, Washington, 1966.

Alan Hirshfeld and Roger W. Sinnott, ed., *Sky Catalogue 2000.0*, Vol. 1, Sky Publishing Corporation, Cambridge, Massachusetts and Cambridge University Press, Cambridge, England, 1982.

Alan Hirshfeld and Roger W. Sinnott, ed., *Sky Catalogue 2000.0* Vol. 2, Sky Publishing Corporation, Cambridge, Massachusetts and Cambridge University Press, Cambridge, England, 1985.

Roger W. Sinnott, ed., *NGC 2000.0,* Sky Publishing Corporation, Cambridge, Massachusetts and Cambridge University Press, Cambridge, England, 1988.

Atlases

Wil Tirion, *Sky Atlas 2000.0*, Deluxe Edition, Sky Publishing Corporation, Cambridge, Massachusetts and Cambridge University Press, Cambridge, England, 1981.

Wil Tirion, Barry Rappaport and George Lovi, *Uranometria 2000.0*, Vol. 1. Willmann-Bell Inc., Richmond, Virginia, 1987.

Wil Tirion, Barry Rappaport and George Lovi, *Uranometria 2000.0*, Vol. 2, Willmann-Bell Inc., Richmond, Virginia, 1988.